BIOLOGICAL CONSEQUENCES
OF
GLOBAL CLIMATE CHANGE

BIOLOGICAL CONSEQUENCES OF GLOBAL CLIMATE CHANGE

Christine A. Ennis
National Oceanic and
Atmospheric Administration

Nancy H. Marcus
Department of Oceanography
Florida State University

UNIVERSITY SCIENCE BOOKS
SAUSALITO, CALIFORNIA

University Science Books
55D Gate Five Road
Sausalito, CA 94965
Fax: (415) 332-5393

Managing Editor: Lucy Warner
Editor: Louise Carroll
NCAR Graphics Team: Justin Kitsutaka, Lee Fortier, Wil Garcia, Barbara Mericle, David McNutt, and Michael Shibao
Cover Design and Photography: Irene Imfeld
Compositor: Archetype Typography, Berkeley, California

This book is printed on acid-free paper.

Copyright © 1996 by University Corporation for Atmospheric Research. All rights reserved.

Reproduction or translation of any part of this work beyond that permitted by Section 107 or 108 of the 1976 United States Copyright Act without the permission of the copyright owner is unlawful. Requests for permission or further information should be addressed to UCAR Communications, Box 3000, Boulder, CO 80307-3000.

Library of Congress Catalog Number: 95-061060

ISBN: 0-935702-85-7

Printed in the United States of America

10 9 8 7 6 5 4 3 2 1

A Note on the Global Change Instruction Program

This series has been designed by college professors to fill an urgent need for interdisciplinary materials on the emerging science of global change. These materials are aimed at undergraduate students not majoring in science. The modular materials can be integrated into a number of existing courses —in earth sciences, biology, physics, astronomy, chemistry, meteorology, and the social sciences. They are written to capture the interest of the student who has little grounding in math and the technical aspects of science but whose intellectual curiosity is piqued by concern for the environment. The material presented here should occupy about two weeks of classroom time.

For a complete list of modules available in the Global Change Instruction Program, contact University Science Books, Sausalito, California, fax (415) 332-5393. Information about the Global Change Instruction Program is also available on the World Wide Web at http://home.ucar.edu/ucargen/education/gcmod/contents.html.

Contents

Preface ix

Introduction 1

I. The Links Between the Biota and Climate 2

The Organization of Living Systems 2
Climate and the Distribution of Living Systems 3
The Importance of Temperature 5
Water and Living Systems 6
Chemical Climate 6
Other Factors 7

II. Climate and the Biota: Forces of Change 9

Increases in CO_2 and Other Greenhouse Gases 9
    Cycling of Carbon Dioxide 9
    Direct Effects of Greenhouse Gases 14
    Indirect Effects of Greenhouse Gases 20

Depletion of Stratospheric Ozone 26
    The Link between the Biosphere and the Ozone Layer 26
    Effects of UV-B on Terrestrial Animals 28
    Effects of UV-B on Terrestrial Plants 30
    Effects of UV-B on Marine Life 32
    Increases in Tropospheric Oxidants 33

III. Uncertainties and Challenges for Future Research 38

Problems and Discussion Questions 40

Appendix 43

Glossary 47

Recommended Reading 51

Index 52

Preface

Scientific discussions of climate change often center on numbers: average temperature increases or decreases to be expected, alterations in rainfall amounts, adjustments to the growing season, and the like. It is easy to lose sight of the fact that we are ultimately interested in climate change because it has the potential to alter the livability of regions and, perhaps, the planet itself. This module steps away from those numbers, the questions of "how much" climate change might occur, and takes a look at the question of "what if." It shows how life forms—plant and animal—might be affected by changes in various aspects of climate. We look at some of the possible biological consequences of the familiar physical aspects of climate change (temperature, rainfall, ultraviolet radiation, etc.). We also explore what we call changes in the "chemical climate," alterations in the chemical composition of the atmosphere that could be substantial enough to elicit responses in biological organisms.

Uncertainties are plentiful in any discussion of biological consequences of climate change, perhaps even more than in other facets of the climate change debate. But that should not deter us from considering the range of possibilities. It is precisely the abundance of uncertainties that makes this one of the most exciting and crucial areas of global change research. Scientists can be sure that every new finding will receive both intense interest and intense scrutiny. This creates a challenge for the scientists and a dilemma for policy makers, who must chart a course for their countries despite the uncertainties. We each must decide how much of a "climate change insurance policy" we will pay for in the face of many unknowns. I hope that this module will help you to answer that question for yourself.

Chris Ennis

Acknowledgments

This instructional module has been produced by the the Global Change Instruction Program of the Advanced Study Program of the National Center for Atmospheric Research, with support from the National Science Foundation. Any opinions, findings, conclusions, or recommendations expressed in this publication are those of the author and do not necessarily reflect the views of the National Science Foundation.

Executive Editors: John W. Firor, John W. Winchester

Global Change Working Group

Louise Carroll, University Corporation for Atmospheric Research
Arthur A. Few, Rice University
John W. Firor, National Center for Atmospheric Research
David W. Fulker, University Corporation for Atmospheric Research
Judith Jacobsen, University of Denver
Lee Kump, Pennsylvania State University
Edward Laws, University of Hawaii
Nancy H. Marcus, Florida State University
Barbara McDonald, National Center for Atmospheric Research
Sharon E. Nicholson, Florida State University
J. Kenneth Osmond, Florida State University
Jozef Pacyna, Norwegian Institute for Air Research
William C. Parker, Florida State University
Glenn E. Shaw, University of Alaska
John L. Streete, Rhodes College
Stanley C. Tyler, University of California, Irvine
Lucy Warner, University Corporation for Atmospheric Research
John W. Winchester, Florida State University

This project was supported, in part, by the
National Science Foundation
Opinions expressed are those of the authors
and not necessarily those of the Foundation

BIOLOGICAL CONSEQUENCES
OF
GLOBAL CLIMATE CHANGE

Introduction

Public awareness of global climate change is extraordinarily high. From schoolchildren to senior citizens, from farms to factories to Wall Street, the phrase "greenhouse effect" now occupies a prominent place in the vocabulary of everyday life. In a world of so many issues and concerns, how is it that global climate change has been able to capture our attention? Why are you interested in the topic?

This problem strikes home for so many people for one simple reason: global climate change has potential consequences for the Earth's living systems (biota). All life on Earth, whether plant or animal, is inexorably linked to the surrounding chemical and physical climate. It is obvious to most people that organisms tolerate certain ranges of temperature, have certain requirements for moisture and nutrients, and respond to the chemical composition of the surrounding air or water. The rate, as well as the magnitude, of global environmental changes could potentially challenge the capability of organisms to adapt. We have a vested interest in this topic because we are members of the biological community ourselves, and because we are dependent on other living systems. It's not surprising that our interest in climate change is intense and that it spans many levels: scientific, economic, social, political, and even emotional.

This module will explore current research concerning the biological consequences of climate change. As with most aspects of climate change, there are few certainties—about exactly how the biota might respond to temperature perturbations, sea-level changes, changes in rainfall patterns, increased ultraviolet radiation, or an altered atmospheric chemical composition. That the range of possible responses is large offers little comfort. In fact, it surely provides even greater motivation to study the topic.

I
The Links Between the Biota and Climate

The rationale for this module is simple: the forces of global change have the potential to alter physical and chemical climate conditions, which determine where organisms can live. In order to understand how climate change might affect living systems, we must first discuss some basics about the biota and how it is linked to climate. Once we understand these links, we will be able to recognize many of the aspects of global change that could spell trouble for the Earth's organisms.

The Organization of Living Systems

A staggering array of life forms makes up the biota of the Earth, and we may approach the study of biological interactions with climate and other features of the abiotic (nonbiological) environment on a variety of levels. Ecologists have come up with conceptual groupings of living organisms. At the simplest level, we talk about individual organisms, composed of cells organized into various tissues and organs that carry out the basic physiological processes of life. The characteristics of the individual are determined by chromosomes, thread-like structures in the nucleus of each cell that carry genetic information and consist of DNA (deoxyribonucleic acid) and proteins. A species is a group of individual organisms sharing a common gene pool. Members of the same species will be similar in appearance, behavior, and chemistry. A population consists of all the members of a species in a particular geographical area at one time. For example, all of the kangaroos in Australia make up a population, which can be described by size, density, age structure, and other properties. The populations of different species in an area are referred to as a community. A community of organisms interacting with one another and with their physical environment constitutes an ecosystem. A field, a lake, and a vacant lot are all examples of ecosystems. A biome is a major type of land community of organisms, such as a desert or a tropical rain forest. Finally, the biosphere is the total of all areas on Earth that support life. It encompasses the deep ocean, land, and the part of the atmosphere (up to an altitude of about eight kilometers) where living organisms are found.

In describing these levels, you can see that we have expanded outward from the very close-up view to the broad view of the globe. That is, the "spatial scale" has gone from millimeters to kilometers to continental to global. The times needed to cause significant change at any level can vary from relatively short, in the case of individuals, to centuries and more at the level of biomes and the biosphere (see Figure 1). When we talk about the effects of climate change on the biota, we could be concerned with any of the above levels of organization. Clearly, studies of effects at the ecosystem or biome level will differ greatly from research at the level of individual organisms because of the differences in temporal (time) and spatial scales. The various types and scales of study are interdependent and complementary in the quest to understand the biotic response to change.

Climate and the Distribution of Living Systems

Living organisms are found everywhere on this planet, even in areas that humans may consider hostile, such as hot springs, the deep sea, and under Antarctic sea ice. However, a given species does not occur everywhere. Species have very specific requirements that govern their distributions. These physical and chemical limits of tolerance, together with availability of food and biological interactions such as predation and competition, determine the global distribution patterns of species. The climate plays a key role in establishing these distribution patterns, insofar as it determines the physical and chemical attributes of the abiotic environment.

Land plants (terrestrial vegetation) offer a good illustration of the connection between climate and the biota. A look at a vegetation map of the world reveals a very close association with the climate zones (tropical, temperate, boreal, etc.; see Figure 2). Both climates and biomes show strong latitudinal zones proceeding from pole to equator. The broad terms for vegetation zones (such as boreal forest, tropical forest, temperate forest) in fact carry climatic

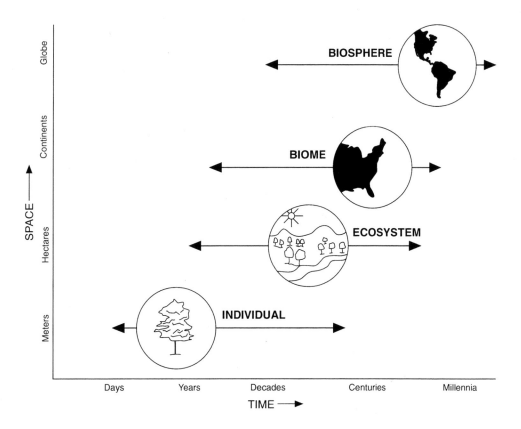

Figure 1. Spatial and temporal scales over which change occurs at various levels of organization of the biota. From Graham, R.L., M.G. Turner, and V.H. Dale, How increasing CO_2 and climate change affect forests, BioScience 40 No. 8, p. 575. Copyright © 1990 by the American Institute of Biological Sciences. Reprinted by permission.

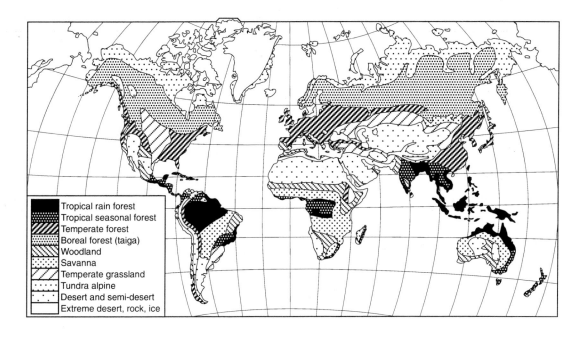

Figure 2a. A map of the world showing where the major biomes occur. From Bolin, B., B.R. Döös, J. Jäger, and R.A. Warrick, eds., SCOPE 29: The Greenhouse Effect, Climatic Change, and Ecosystems, *John Wiley & Sons. Copyright ©1986 by the Scientific Committee on Problems of the Environment (SCOPE). Reprinted by permission.*

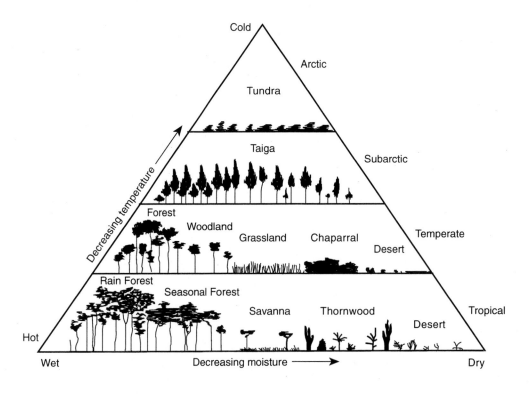

Figure 2b. The relationship between climate and biome. This diagram shows that temperature and precipitation are the main factors determining what biome occurs where. From Arms, K., Environmental Science. *Copyright © 1990 by Saunders College Publishing. Reprinted by permission of the publisher.*

connotations. This offers the first clue that climate variables influence plant distribution. Likewise, the fossil record provides evidence of climatic factors acting over millions of years to control species distributions. Modern science can hardly lay claim to the discovery of this basic relationship; it was recognized by Greek scholars between the third and fifth centuries B.C. One emphasis of contemporary research is to go beyond the fundamental climate-distribution correlation in an effort to understand the mechanisms by which climate exerts its control.

What are some of the fundamental physiological and ecological principles that link living systems so strongly to climate factors?

The Importance of Temperature

Temperature is the most important climatic factor that governs the biology of animals and plants. It has a direct effect on the rate of most biological processes (e.g., photosynthesis, respiration, digestion, excretion). If proper internal temperatures are not maintained, these processes cannot proceed normally and an individual will be stressed or possibly die. You may wonder why temperature is so critical. It's because physiological processes are catalyzed (that is, their rate is determined) by a group of molecules known as enzymes. The structure of these enzymes is dependent on temperature; if temperature is too high, the enzyme molecule is rearranged ("denatured"), and it simply doesn't work the way it should. If temperature is too low, the enzyme isn't effective at catalyzing physiological processes, and those processes slow down. Thus chemistry, or more specifically biochemistry, is behind the basic temperature-tolerance limits of organisms. Figure 3a illustrates in a general way how an organism's functioning may respond to environmental parameters such as temperature. Figure 3b shows the specific case of the response of plant photosynthesis (the process whereby plants use sunlight, carbon dioxide, and water to make carbohydrate food) to temperature.

Species differ in their abilities to regulate body temperature. The internal temperature of "poikilothermic" organisms (e.g., insects and reptiles) tends to vary over a much greater range than that of warm-blooded or "homeothermic" organisms (e.g., mammals and birds), which are able to regulate their body temperature by internal physiological mechanisms. For example, mammals maintain a body

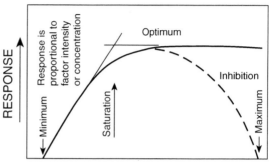

Figure 3a. A generalized curve showing organism response to an environmental parameter. Minimum, optimum, and maximum are called out as cardinal points. From Salisbury, F.B., and C.W. Ross, Plant Physiology. *Copyright © 1985, 1978 by Wadsworth, Inc. Reprinted by permission.*

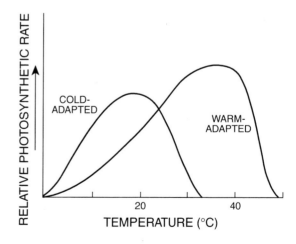

Figure 3b. The dependence of net photosynthetic rate on temperature for cold- and warm-adapted plants. From Gates, D.M., Biophysical Ecology. *Copyright © 1980, Springer-Verlag. Reprinted by permission.*

temperature of approximately 35–39° C and birds 38–42° C. The mechanisms for controlling internal temperature, such as shivering, sweating, and panting, work by altering the organism's metabolic rate. Because warm-blooded animals are able to regulate their body temperature, they are less affected by external temperature variation. Poikilothermic organisms are able to achieve some degree of temperature regulation by modifying their behavior (e.g., lizards bask in the sun). Many cold-blooded organisms are dark colored, an adaptation that allows them to directly absorb more solar radiation. (Note that this links living organisms to another climate factor, sunshine, which is not the same as temperature.)

Several aspects of temperature may limit a species. In some cases it may be the maximum temperature, in others the minimum, the average, or the frequency with which temperature changes. If environmental temperatures rise, those species living near the upper thermal limits of their existence will be stressed. Although death may not be the immediate response, sublethal effects such as a decrease in the number of offspring may contribute to the eventual decline or demise of the species.

Water and Living Systems

Moisture is also a critical variable limiting the distribution of organisms. Water, which accounts for 85 to 90% of the weight of most living organisms, is essential for proper plant and animal physiology. This is because the chemistry of physiological processes is carried out in the water of living tissues. Water is the equivalent of an organism's transportation system, permitting the transport of biological molecules to and from the sites of biochemical reactions such as photosynthesis and respiration. Nevertheless, organisms differ greatly in their water requirements and their abilities to withstand dryness and flooding. Some lower plants, such as algae and fungi, adjust their activity according to the humidity of the surrounding air. In periods of low humidity, they are able to enter a resting state. Some water-dwelling microorganisms have a similar ability to withstand periods of total dryness.

This dependence on water links the biota to several climate-related variables, such as relative humidity, rainfall amount, and the distribution of rainfall through the year. Figure 4 shows how plant growth is related to a very simple climate variable, total yearly precipitation. The frequency of extreme events such as droughts and floods is also critical. Reproduction is especially vulnerable to such extremes, with the survival of eggs and seeds often tied to the presence or absence of moisture. For example, amphibians such as toads and salamanders lay eggs in the fringes of ponds, placing them at risk in periods of drought. Larger animals can sometimes move (migrate) to minimize the impacts of extreme events, but plants and smaller animals don't have this option. The timing and/or duration of rainfall are other critical climate features for some organisms.

Chemical Climate

All organisms are sensitive to the chemical composition of the medium they live in, whether that medium is air or water.

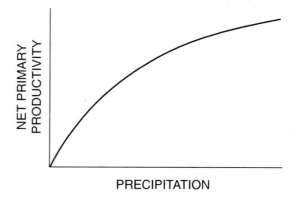

Figure 4. Idealized curve showing the dependence of plant growth on total annual precipitation.

Organisms require certain chemicals, oxygen being the most obvious example for all but the anaerobic microorganisms. In the case of plants, carbon dioxide (CO_2) is required.

Beyond these essential needs, other chemicals can only be tolerated at certain levels. For water-dwelling (aquatic) species, the pH (acidity) of the water is a critical variable. Acid rain, currently a problem in some regions, could alter the pH of lakes or coastal areas enough to affect the health and survival of aquatic plants and animals. Pollution of streams or coastal waters by toxic organic chemicals and metals such as mercury and lead also has biological consequences. Water pollution can present serious local problems, as in Mexico City, where water has been rendered unfit for human consumption. While these are serious environmental problems, we shall not be discussing them further because the focus of the module is on problems that are global in their scope.

Likewise, organisms that depend on air have limited tolerance for some atmospheric gases (Table 1). Modern-day air pollutants like sulfur dioxide (SO_2) and ozone (O_3) affect plants and animals by interfering with basic physiological processes such as photosynthesis and respiration. These gases are the by-products of industrial processes that have increased as the human standard of living has advanced. Some toxic gases such as phosgene have been developed deliberately as chemical weapons because of their rapid and serious disruptions to human physiology. In this module, we will discuss one aspect of air composition—the occurrence of gaseous oxidants—which may be becoming a problem of global dimensions.

Other Factors

A few other climate factors also affect living systems. While the activities of humans do not affect the amount of solar radiation striking the

Table 1
Major Air Pollutants and Effects

Pollutant	Principal adverse effects
Particulate matter	Increased respiratory illness and mortality; visibility impairment; materials damage
Sulfur dioxide*	Increased respiratory illness and mortality; plant and forest damage; materials damage
Carbon Monoxide	Asphyxiation; mortality in cigarette smokers; impaired functioning of heart patients
Ozone	Increased respiratory symptoms and illness; plant and forest damage; materials damage
Nitrogen oxides*	Increased respiratory illness; leads to ozone formation; materials damage
Lead	Neurological malfunctioning; learning disabilities; increased blood pressure

*Results in acid rain which causes health, aquatic, plant, and materials damage.

Atmospheric chemical factors that are harmful to the biota. From Cortese, A.D., Clearing the air. Environmental Science & Technology 24, No. 4. Copyright © 1990 by the American Chemical Society. Reprinted by permission.

outer atmosphere of the Earth, our activities can affect cloudiness and thus the amount of sunlight that reaches the surface of the Earth. Solar radiation is important for many organisms, most obviously plants, which require light for photosynthesis. Visible light in the range of 400–700 nanometers in wavelength, called "photosynthetically active radiation," or PAR, is captured by chlorophyll and other pigment molecules in plant cells and eventually is converted to chemical energy and stored in the molecular products of photosynthesis. (The chemistry of photosynthesis is explained on pages 9–10). Plants can get too much of a good thing, however. Thin-barked trees such as aspen and birch can suffer from sunscald. Sunshine also plays a role in the distribution of some animals. One ant species requires a habitat with at least 40 days per year of full sunshine. Some reptiles, known as heliotherms, depend on solar radiation to regulate body temperature. While many organisms require visible light, they can only tolerate certain levels of other components of solar radiation. Ultraviolet light of some wavelengths is actually harmful to many plants and animals, as we shall discuss later in this module.

Atmospheric and oceanic circulation are other features of climate that have a bearing on living systems. Wind seems to be important as a means of transport for some insects such as moths, dragonflies, and locusts. The location of the major prevailing winds often determines the distribution of migratory birds, which may use the winds quite deliberately to assist their travels. This strategy occasionally backfires, as winds unexpectedly carry the birds farther than their food reserves allow. Wind is an important mechanism for seed dispersal of some plants, which have evolved seed casings that capitalize on the free ride. (Dandelions and maple trees are familiar examples.) But one species' gain can be another species' loss: windy conditions can be deadly to young insects and harmful to coastal-dwelling plants and animals. Oceanic currents can also be either a help or a hindrance to organisms. Phytoplankton and zooplankton, including the immature stages of some bottom-dwelling marine animals, are often carried long distances by oceanic currents. This can be a mechanism to extend the geographical range of a species. But individual organisms will die if they are transported to regions outside their climatic tolerance levels.

II
Climate and the Biota: Forces of Change

Within any given species, the abilities to acclimate or migrate provide possible means of coping with new climatic conditions. There also may be large genetic variation within a given species, increasing the chance that some individuals will survive environmental change. If organisms do not have these abilities to cope with change, the deficiency could prove fatal. The range of possible responses is large. In the rest of this section, we'll explore some of scientists' best guesses about how global changes could impact the biological world.

Although humans influence the Earth in myriad ways, there are only a few examples that are truly global in their scope. This module will focus on three cases that have implications for living organisms. Two are likely to be very familiar: increases in greenhouse gases, such as carbon dioxide, and stratospheric ozone depletion. The third topic is less familiar because it has only recently been considered a global problem: the increase in oxidants (such as ozone) in the lower region of the atmosphere known as the troposphere. These seemingly diverse changes actually have two fundamental characteristics in common. First, each of these changes is being brought on by chemicals that we are adding to our atmosphere (carbon dioxide, chlorofluorocarbons, nitrogen oxides, volatile organic compounds). Second, each of these human-induced, or anthropogenic, changes has the potential to affect the fundamental physiological functions of life forms, either directly or indirectly, in one or more ways.

Increases in CO_2 and Other Greenhouse Gases

Cycling of Carbon Dioxide

Carbon dioxide (CO_2) is an important molecule in the basic physiological processes carried out by living organisms. Respiration, occurring in your body right now (we hope!), involves the consumption of oxygen and the release of CO_2 in a series of chemical reactions. The ultimate benefit of this process to you, as an organism, is that it allows you to break the chemical bonds in food molecules to release energy. This energy can then be used in numerous ways, to build cells, move muscles, repair injury, combat infections, etc. Both animals and plants carry out respiration as an energy-releasing mechanism.

Plants, including marine algae, can turn the tables with respect to CO_2. They also consume CO_2 in the process of photosynthesis (see Figure 5). This process, which is initiated by light, converts CO_2 and water (H_2O) in a series of chemical reactions that release oxygen (O_2) and produce various organic substances. The organic substances make up the "stuff" of plant matter, including the stems, roots, leaves, and flowers. In a real sense they are the "stuff" of animal matter, too, since plants are the ultimate food source of all animals, even carnivores. (Thus, plants are called the "primary producers.") Figure 6 gives examples of some of the wondrously complicated organic materials that

PHOTOSYNTHESIS (ENERGY-STORING PROCESS)

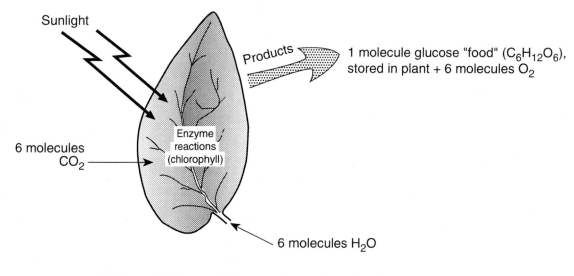

$$6\ CO_2 + 6\ H_2O + \text{sunlight} \longrightarrow 1\ C_6H_{12}O_6\ (\text{glucose}) + 6\ O_2$$

RESPIRATION (ENERGY-RELEASING PROCESS)

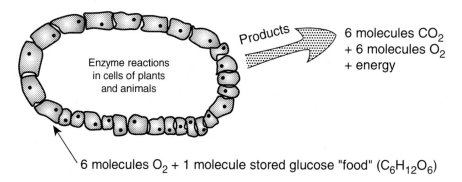

$$6\ O_2 + 1\ C_6H_{12}O_6\ (\text{glucose}) \longrightarrow 6\ CO_2 + 6\ O_2 + \text{Energy}$$

Figure 5. A simplification of the chemical processes of respiration and photosynthesis. Carbohydrates, shown here using the example of glucose, are produced in photosynthesis. The stored chemical energy of carbohydrates is released during respiration. Photosynthesis occurs only in plants, but respiration occurs in plants and animals. Although these two overall processes are nearly the reverse of each other, the individual detailed steps involved in each process are very different.

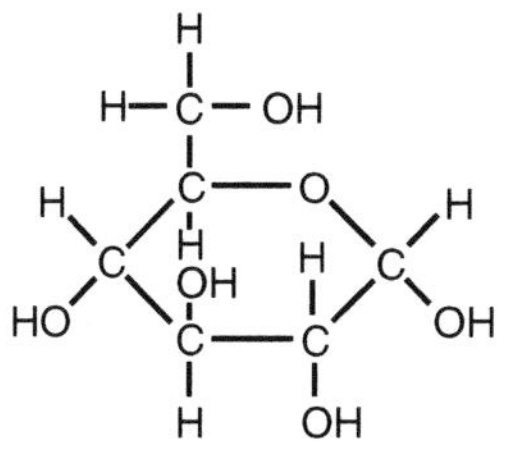

Ring form

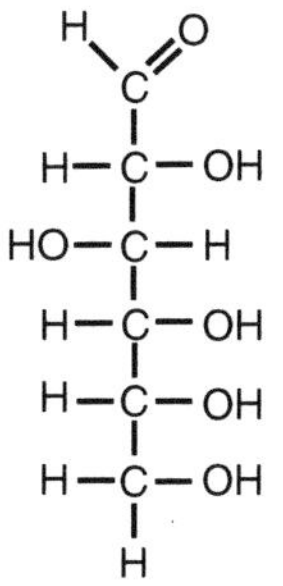

Open-chain form

Glucose, $C_6H_{12}O_6$

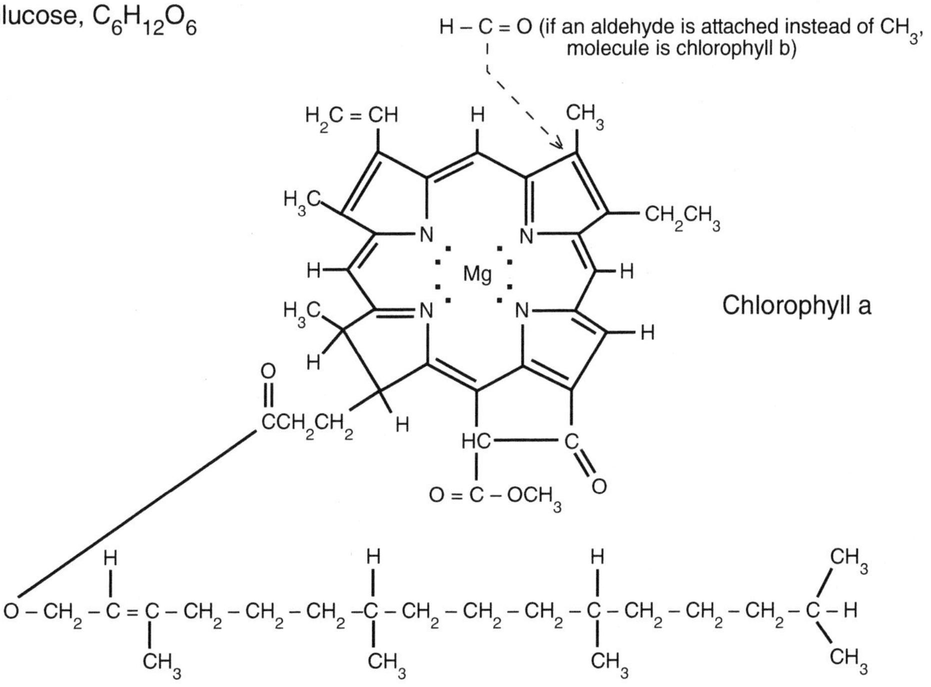

Chlorophyll a

phytol tail ($C_{20}H_{39}$)

Figure 6. Examples of molecules that make up plant matter. The vertices of the ring-like structures are carbon atoms, which are not drawn as "C" in order to keep the diagram readable. Photosynthesis has captured all of these carbon atoms from atmospheric CO_2 molecules. Various chemical processes in the plant then lead to the production of these wondrously complex organic substances. From Salisbury, F.B., and C.W. Ross, Plant Physiology. *Copyright © 1985, 1978 by Wadsworth, Inc. Reprinted by permission.*

a plant makes from the simple starting materials of light, CO_2, and water. Note from Figure 5 that photosynthesis is the reverse of respiration with respect to O_2/CO_2 production and consumption.

The important point about the processes of respiration and photosynthesis is that the carbon atoms are being "cycled" from one form (CO_2 in the atmosphere) to another (organic molecules in plant and animal tissues) and then back again, through processes such as respiration and decomposition. The biosphere and atmosphere are communicating, via the CO_2 molecule. This and other aspects of the global "carbon cycle" are represented schematically in Figure 7. This figure also shows two ways in which humans have perturbed this cycle of storage and release. The consumption of organic fuels (oil, coal, gas, wood) and the process of deforestation essentially unlock carbon atoms stored in the fuel/forests and release them to the atmosphere as CO_2 during the burning process. In so doing, humans are short-circuiting the normal storage times for carbon in these forms. The amount of CO_2 in the atmosphere would be increased, assuming that the processes that remove it from the atmosphere (such as being dissolved in the ocean or absorbed by plants through photosynthesis) cannot keep pace.

In fact, there is now indisputable evidence that global levels of carbon dioxide are increasing in the Earth's atmosphere. This trend can be seen in the now-famous record of CO_2 observations from the Mauna Loa Observatory in Hawaii. Figure 8 shows the monthly concentrations of CO_2 for the last few decades. Besides the overall increase in CO_2, it is interesting to see the peaks and valleys that occur within each year. Summertime brings on the valleys, because the terrestrial plants are active and photosynthesis is removing CO_2 from the atmosphere. As the growing season progresses to fall and winter, plants are less active and atmospheric CO_2 increases. Figure 8 thus shows the biosphere "breathing." This can be seen on a larger scale, indirectly, through satellite images that portray the "greenness" of vegetation. Outside of the tropics, the greenness can be seen to wax and wane during a year, with the northern and southern hemispheric cycles opposing one another in an elegant symmetry.

One reason the increase in CO_2 is important is because CO_2 is a greenhouse gas. Its molecular structure enables it to absorb infrared radiation that would otherwise escape to space, trapping it like the windowpanes on a greenhouse do. Not all molecules have a structure that permits the absorption of significant amounts of infrared radiation, so the greenhouse gases are like an elite club. Other members of the club are methane (CH_4), nitrous oxide (N_2O), chlorofluorocarbons (commonly referred to as freons), and ozone (O_3). Like CO_2, the amounts of all of these greenhouse gases are increasing on a global scale. To a first approximation, we might guess that increases in greenhouse gases might lead to a warming of the planet (hence the term "global warming"). But there are many feedbacks in the Earth system, and it is a complicated problem that has implications for more than just temperature. We will discuss these global implications later, but first we'll look at the ways in which the biota may be affected more directly by increases in greenhouse gases.

CLIMATE AND THE BIOTA: FORCES OF CHANGE

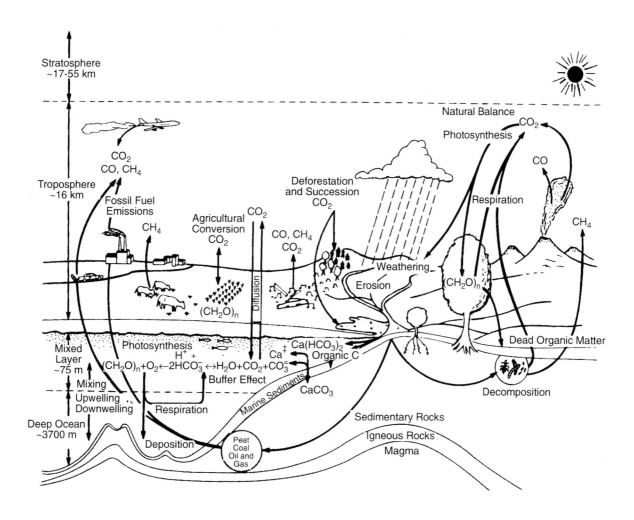

Figure 7. The carbon cycle. The atmosphere and oceans are the Earth's greatest storage tanks of carbon, with a sizeable amount residing in carbonate rock and fossil fuels. The cycling of carbon begins with atmospheric CO_2 entering plants during photosynthesis and becoming incorporated into carbohydrates [$(CH_2O)_n$]. Plants then carry on respiration, releasing some of the carbon back into the atmosphere and soil as CO_2. Some of the plant carbohydrates are eaten by animals, where they can be stored or where respiration can return them to the atmosphere as CO_2. The complete cycle requires the process of decomposition, whereby microorganisms break down dead plant and animal matter in their own respiration to return the carbon to the atmosphere as CO_2. Methane (CH_4) and carbon monoxide (CO) are also important carbon-containing gases. They have natural and human sources as shown. Two significant sources of CO_2—the burning of fossil fuels and of tropical forests by humans—have recently been added to the carbon cycle. From Trabalka, J.R., ed., Atmospheric Carbon Dioxide and the Global Carbon Cycle, 1985. Washington, D.C.: U.S. Department of Energy.

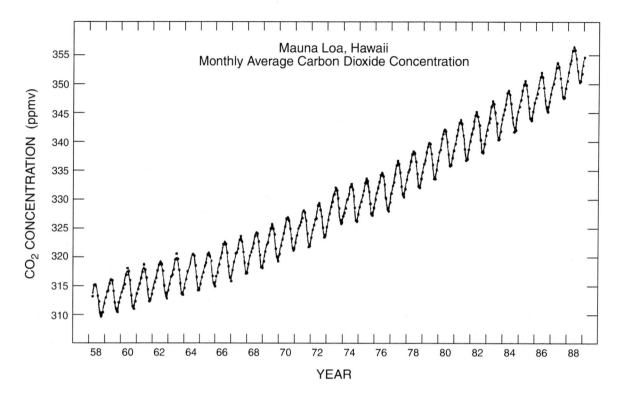

Figure 8. Variation of atmospheric concentrations of carbon dioxide at Mauna Loa Observatory, Hawaii. Based on data from C. Keeling, Scripps Institution of Oceanography.

Direct Effects of Greenhouse Gases

Because carbon dioxide is essentially a food material for plants, it is logical to expect that the increase in atmospheric CO_2 might have some direct effects on vegetation. Much research into this possibility is now occurring. On the level of individual plants, photosynthesis and growth are often stimulated by higher concentrations of atmospheric CO_2. This effect is known as "CO_2 fertilization," and plant growers have known about it for some time. In fact, the air in greenhouses is often spiked with higher CO_2 for this very reason. Figure 9 shows some sample data for photosynthesis in soybeans.

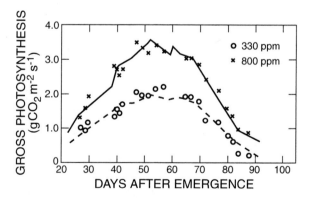

Figure 9. Photosynthetic rate for soybean plants grown at two CO_2 concentrations. Present-day CO_2 is about 350 parts per million (ppm). From Strain, B.R., and J.D. Cure, Direct Effects of Increasing Carbon Dioxide on Vegetation, 1985. Washington, D.C.: U.S. Department of Energy.

An interesting side effect has been observed for plants that are grown in high-CO_2 conditions: they are able to use water more efficiently. This is because they can get sufficient CO_2 without opening their pores or "stomates" as widely (see Figures 10a and 10b). This leads to less water loss from the interior of the plant. Just as water evaporates more slowly from a narrow-necked container than it does from a wide-mouthed container, it evaporates more slowly from a closed stomate. For every CO_2 molecule that enters the stomates, 100 to 400 water molecules can be lost from inside the leaf. So, reductions in stomatal openings can lead to significant water savings.

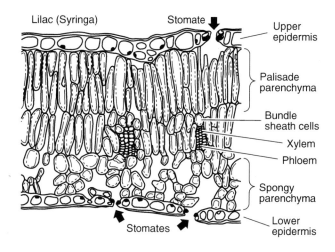

Figure 10a. Cross section of a lilac leaf, showing various types of cells and the stomatal openings. From Salisbury, F.B., and C.W. Ross, Plant Physiology. *Copyright © 1985, 1978 by Wadsworth, Inc. Reprinted by permission.*

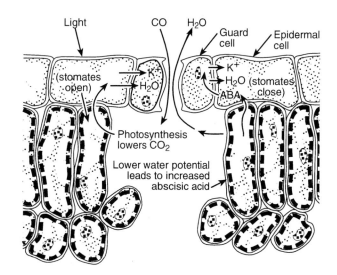

Figure 10b. Much like balloons, the "guard cells" of plants inflate and deflate with water to open and close the stomatal pores. This figure shows two important feedback loops, one for CO_2 and one for H_2O, that control stomatal action. The left part of the drawing illustrates the effects of light. Light promotes photosynthesis, which lowers CO_2 levels in the leaf; the leaf's response is to cause more potassium ions (K^+) to move into guard cells, and water follows osmotically, causing stomates to open. (There is also a direct effect of light on stomatal opening, independent of CO_2 levels.) The right-hand side shows the effects of water stress. When more water exits the leaf than can enter from the roots, abscisic acid (ABA) is released or produced from nearby cells, which leads to movement of K^+ out of guard cells; water follows osmotically, so stomates close. If the rate of drying is extremely rapid, water is lost from the guard cells directly, bypassing the ABA step but still leading to closure. From Salisbury, F.B., and C.W. Ross, Plant Physiology. *Copyright © 1985, 1978 by Wadsworth, Inc. Copyright © 1969 by Wadsworth Publishing Company, Inc. Reprinted by permission.*

Some plants might respond to increased CO_2 differently from others because of differences in their basic machinery of photosynthesis. Studies of photosynthetic pathways have resulted in the classification of plants into two major groups, C_3 and C_4, based upon the number of carbon atoms in the compound that is synthesized during the early steps of photosynthesis (see Figure 11). These two major groups occur in both natural and managed (agricultural) systems, but their prevalences vary from ecosystem to ecosystem. Agricultural crops include C_3 plants (wheat, rye, barley, oats, soybean, sunflower, rice, alfalfa) and C_4 plants

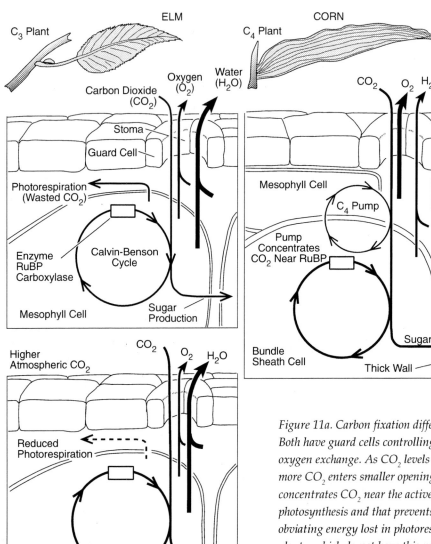

Figure 11a. Carbon fixation differs between C_3 and C_4 plants. Both have guard cells controlling carbon dioxide, water, and oxygen exchange. As CO_2 levels rise, both lose less water as more CO_2 enters smaller openings. C_4 plants have a pump that concentrates CO_2 near the active site of an enzyme crucial to photosynthesis and that prevents oxygen from binding—obviating energy lost in photorespiration. In high CO_2, C_3 plants, which do not have this pump, find CO_2 naturally concentrated at the site and do not lose energy by binding oxygen. From Bazzaz, F.A., and E.D. Fajer, Plant life in a CO_2-rich world. Scientific American, January 1992. Original figure by Patricia J. Wynne, Copyright © 1992 by Scientific American, Inc. Reprinted by permission.

(corn, sugar cane, sorghum, millet). Most trees are C_3 plants, but many tropical plants are C_4. C_4 plants appear to have originated in the tropics, in fact. They are more tolerant of, and grow very well at, higher temperatures, low water levels, and high light intensities. Detailed studies have revealed why this might be so. These plants have a mechanism whereby they can build up CO_2 to higher concentrations within their leaves. Because of this, they can function without opening their stomates as much as the C_3 plants. Because the C_4 plants already hold high levels of CO_2 within their leaves, researchers think that they might not be fertilized by additional CO_2 as much as the C_3 plants.

This last point leads to an interesting possibility. Enhanced CO_2 could lead to shifts in the mixture of species within communities. If the performance of C_3 plants improves more than that of C_4 plants, the C_3 plants could gain a competitive edge that enables them to become more prevalent. Indeed, studies have demonstrated this effect (see Figure 12).

The reproduction and flowering of plants might be affected by carbon dioxide concentrations in the atmosphere. Studies of several species have demonstrated that flowering time is advanced at higher carbon dioxide levels. In addition, several floral and reproductive characteristics may be modified, including the number of flowers, the rate of flower production, the longevity of flowers, and seed weight. These changes may have a dramatic impact on the survival of insect-pollinated species. The relationship between a plant and the insects that

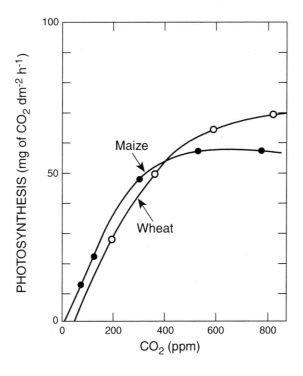

Figure 11b. Response of photosynthesis to CO_2 concentration for a C_3 species (wheat) and a C_4 species (maize). From Bolin, B., B.R. Döös, J. Jäger, and R.A. Warrick, eds., SCOPE 29: The Greenhouse Effect, Climatic Change, and Ecosystems, John Wiley & Sons. Copyright © 1986 by the Scientific Committee on Problems of the Environment (SCOPE). Reprinted by permission.

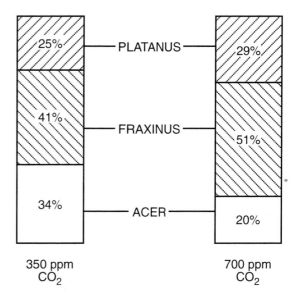

Figure 12. Change in species composition of a tree community in response to CO_2 at two concentrations. Numbers are percent of total biomass. Adapted from the data in Bazzaz, F.A., K. Garbutt, and W.G. Williams, 1985. The Effects of Elevated Atmospheric CO_2 on Plant Communities (DOE/EV/04329-5). Washington, D.C.: U.S. Department of Energy.

assist in pollination is the product of a long coevolutionary process and is highly specific and often complex. If the insect pollinator cannot adjust to changes in the plant such as flowering time, reproduction of the plant could be jeopardized.

Nearly all of the research on this topic has been carried out using greenhouse-grown plants, in managed environments, so we really aren't sure how plants in the natural environment might respond. Do they adjust to gradual changes in CO_2 during their lifetime? Some results suggest that they do. Recent experiments show that plants that are grown in high CO_2 sometimes behave quite differently from plants that are grown in one CO_2 level and then suddenly exposed to higher CO_2 in an experiment. Some of the gains in growth are not observed when the experiment is conducted gradually and over the longer term. One suggested explanation is that the plant's ability to make photosynthetic products (carbohydrates) outruns its ability to move those products out of the leaf to other parts of the plant. The plant then slows down its photosynthesis through a biochemical feedback. Another possible explanation is that supplies of the nutrient phosphorus and/or enzymes needed in photosynthesis are unable to keep pace with a higher rate of photosynthesis.

There are other questions to be answered. Some recent research suggests that CO_2 fertilization may occur only when plants have adequate supplies of water and nutrients such as nitrogen, phosphorus, and sulfur (see Figure 13). Since most of the early research was done on plants that were fertilized and watered regularly, the plants were able to utilize the additional CO_2. But in natural systems, the reality is that nutrients and/or water often limit growth, so the plant cannot take advantage of extra CO_2. Soil properties, such as the ability of the soil to retain water, could also be a limitation. This raises a major uncertainty: on what scale can we draw conclusions from the existing body of scientific research? Even if studies show that individual plants respond positively to higher CO_2, entire plant communities might not show enhanced growth. There is little information about how whole ecosystems might respond to increased CO_2. Two long-term studies on ecosystems have had contradictory results. Arctic tundra ecosystems appear to have species and environmental limitations that prevent them from responding to elevated CO_2, whereas a salt marsh dominated by a C_3 species did show an increase in productivity. In each case, nutrient availability may explain the results. In the tundra, nutrients may have been inaccessible to the plants because they were frozen in the permafrost. In the marshland, nutrients and water were abundant and available.

Conducting experiments on the scale of whole ecosystems is quite difficult, and yet it is crucial for us to be able to understand the kinds of changes that might occur on this scale. Computer models of ecosystems are one tool that has been used to try to extend our understanding to larger temporal and spatial scales. These models must ultimately rely on experimental information for their input, but even with this limitation they can be particularly useful as a

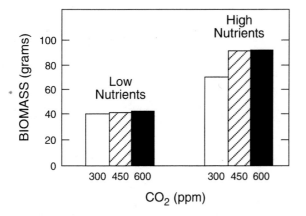

Figure 13. Response of a community of plants to "CO_2 fertilization" at two different levels of soil nutrients. Adapted from Zangerl, A.R., and F.A. Bazzaz, 1984. The response of plants to elevated CO_2 II. Competitive interactions between plants under varying light and nutrients. Oecologia 62: 412–17.

way to identify which processes need further research.

Despite major uncertainties, the topic of CO_2 fertilization is important because it is a possible "feedback" between climate change and the biosphere. The hypothesis is that increased atmospheric CO_2 would stimulate photosynthesis, increased photosynthesis would consume greater amounts of atmospheric CO_2 than would otherwise occur, and atmospheric CO_2 levels would then decrease. In other words, the plants would be acting like a sponge, mopping up some amount of the excess CO_2 that humans are adding to the atmosphere (see Figure 14). If true, this would be what is known as a "negative feedback" (the result is counter to the initial event and is therefore stabilizing). There is much debate now in the scientific community as to whether CO_2 fertilization occurs in the real world and whether it could be significant enough to buffer some of the additional CO_2 that will be added to the atmosphere in the future.

We have described much research about the response of plants to increases in CO_2. What about animals? Many of the possible responses of plant communities described in the previous several paragraphs could have effects on the animals that depend on them for food, shelter, or mating sites. Changes in species composition, or a general decline in species diversity, could undermine the health of natural ecosystems. Another possible ecosystem-level impact comes from the finding that plant material (leaves, stems, etc.) contains less nitrogen and protein when the plant is grown in a CO_2-rich environment. We don't yet know why this is so, but it means that the plant material becomes less nutritive to insects and other animals that feed on it (herbivores). To compensate for the decreased protein content of the plant, herbivores may either feed more or grow less. Hence, although the photosynthetic rate of a plant may increase due to the direct effects of CO_2 fertilization, increases in herbivore grazing may offset any gains in growth. Possible impacts on herbivores complicate the picture even further. For example, research shows that the early developmental stages of the buckeye butterfly are slowed when the butterfly must feed on poorer-quality plants. This could translate into population declines for this insect and eventually affect organisms higher up in the food chain.

Changes in carbon dioxide also might have an impact on marine life. In the marine system, CO_2 reacts with water to form bicarbonate and carbonate ions. These substances play a major role in the oceans by stabilizing the pH of seawater at approximately 8.0–8.2 (pH values above 7 are basic and below 7 are acidic). Because of the intimate association of CO_2 and pH, scientists predict that a doubling of carbon dioxide in the atmosphere will lead to a de-

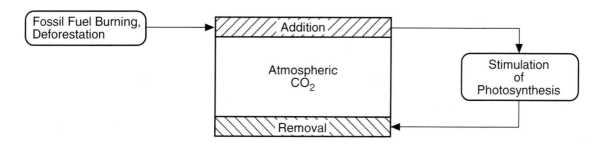

Figure 14. Example of the CO_2 fertilization feedback. Additional CO_2 in the atmosphere stimulates plants to consume more CO_2 in photosynthesis, thus removing CO_2 from the atmosphere. This feedback is stabilizing (negative).

crease in the oceans' pH by 0.3 units, making the oceans more acidic than at present. This seemingly minor change will greatly alter the concentration of essential nutrients and trace metals that are required by phytoplankton (the microscopic plants of the sea) for the process of photosynthesis. A change in phytoplankton photosynthetic rates ripples upward to higher levels in the food chain, which all ultimately depend on these primary producers for their food (see Figure 15). The fishing industry is all too familiar with this phenomenon; it occurs periodically when a natural occurrence known as El Niño alters the supply of nutrients in the surface waters in some areas of the world's oceans. (El Niño is the subject of another module in this series.)

The greenhouse gases methane (CH_4), nitrous oxide (N_2O), and chlorofluorocarbons (CFCs) are not expected to have direct effects on biota. However ozone (O_3), another greenhouse gas, does have serious direct impacts on plants and animals. In fact O_3 is different from the other greenhouse gases because its effects on biological organisms, and its importance in the chemistry of the lower atmosphere (troposphere), are due to the fact that it reacts so readily with other molecules. For this reason, we have separated ozone from the other greenhouse gases, and we shall discuss its biological effects in a separate section later in the module.

Indirect Effects of Greenhouse Gases

The increase in CO_2 and other greenhouse gases is expected to alter the physical climate, which will, in turn, impact living organisms in various ways. We call these "indirect" effects, because the greenhouse gas molecule itself is not what the biota is responding to.

Global climate models now indicate that temperature and rainfall are likely to change as a result of the buildup of greenhouse gases. Both increases and decreases are possible, depending on which region of the world we are considering. Precipitation may increase at high latitudes, while the middle latitudes of the Northern Hemisphere may experience less summer rainfall and soil moisture. Not only averages but annual patterns of temperature and precipitation may change. As we discussed

Figure 15. The food pyramid of the ocean. The tiny phytoplankton at the top capture the energy of the sun. These are eaten by animals larger than themselves, which are, in turn, eaten by larger animals. At the bottom are the largest carnivores of the sea. (There are no large herbivores in the open oceans as there are on land.) Note that each level is composed of far fewer organisms than the one above it. The levels are purely theoretical; creatures at the top of the food chain may actually live in the deep ocean, and vice versa. From Wallace, R.A., J.L. King, and G.P. Sanders, Biology: The Science of Life, *Copyright © 1981 Scott, Foresman and Company. Reprinted by permission of Harper Collins College Publishers.*

in Section 1, these are important aspects of the physical climate, which we would expect to impact living systems.

The C_3 and C_4 plants might respond to altered temperature and rainfall differently because of the differences in their basic machinery of photosynthesis. As we noted in the previous section, C_4 plants are more tolerant of, and grow very well at, higher temperatures, low water levels, and high light intensities, because they can concentrate CO_2 within their leaves and operate with smaller openings in their pores. This allows them to retain water better. Kentucky bluegrass and creeping bent, both C_3 species, are common lawn grasses in cooler parts of the United States. During the summer months, however, they often become overgrown by crabgrass, a C_4 species that fares better under hotter conditions. A first guess is that C_4 plants may have the upper hand in areas of increasing temperature and decreasing moisture.

A 1°C increase in average temperature may seem inconsequential. In terms of climate, this change corresponds to a 100–150 kilometer distance in latitude! How will plants and animals adjust as their optimal distribution ranges literally shift out from under them? A very important area of research deals with the complex issues of whether species will migrate as their optimal temperature and moisture ranges shift. Some larger animals may be capable of migrating, but smaller animals and plants are much more limited. Tree species do migrate gradually to follow favorable environments during climate change. In the Ice Age, for example, tree species in Europe and North America migrated at an average rate of 300 meters per year. It is hard to say whether global warming would evoke a similar response, because the predicted rate of change is 10 to 40 times faster than with any previous climate change. In addition, the presence of humans distinguishes this warming period from its predecessors. Humans have divided the landscape and fractured the continuous ranges of many organisms. As a result, the migration routes of some organisms will be hindered by cities, roads, and farmland. On the other hand, humans might intervene to assist the migration of trees, other vegetation, and animals through programs of conservation, transplantation, or relocation.

Some researchers predict major changes in the forests of North America. They expect that the distribution of the four dominant tree species (yellow birch, sugar maple, hemlock, and beech) will be shifted 500 to 1,000 kilometers northward and that all four species will become extinct along their present-day southern borders. Coincident with the shift of the southern border northward will be an increase in suitable habitat along the northern border by as much as

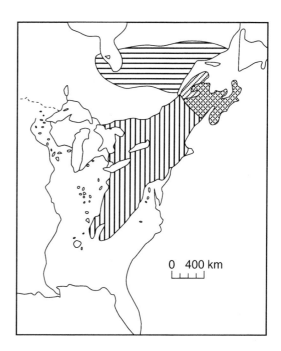

Figure 16. Present and future range for a common forest tree, eastern hemlock (Tsuga canadensis), under a doubling CO_2 climate scenario. Horizontal shading is the present range, and vertical shading the potential range with CO_2 doubling. Cross-hatched area of overlap is where the trees are likely to be found 100 years from now. Relict colonies may persist to the south in pockets of favorable environments. From IGBP Report No. 12 (1990), National Research Council, 1986. Reprinted by permission.

500 kilometers. See Figure 16 for the predicted range of one species. Although temperatures might be suitable for survival at the more northern latitudes, it is not clear that the trees will be able to migrate (via the dispersal of pollen) at a rate fast enough to keep pace with the temperature change. For example, pollen data indicate that beech trees can migrate about 20 kilometers per century, which falls far short of the rate of predicted shifts in climate zones (about 200–500 kilometers per century). The mixture of species in communities could change, as different species move at different rates and have different lag times in their responses to climate change.

Genetics adds another layer of complication to this problem. It could be that the trees in northern areas have evolved in response to the colder conditions and would not be able to acclimate to warmer temperatures, even if those warmer temperatures were in the tolerance range of the species in general. For example, the beech trees in Maine may be genetically distinct from the beech trees in Georgia, but transplantation of southern trees to northern locations might be the only means to ensure survival of the species. This example illustrates the importance of genetic variation within a species. If individuals within a species are genetically very different from one another, the capacity of the species to respond to and survive change is greater than that of a species of individuals that are all alike. The diversity in the former case increases the likelihood that at least some individuals will be able to tolerate the new conditions.

Temperature change could have repercussions for the process of decomposition. In this process organisms such as bacteria and fungi feed on dead plant or animal material in soils, accomplishing the important task of releasing the nutrients contained in the material (see Figure 17). Decomposition rates accelerate as temperature increases, but are slowed by dryness. Changes in plant nitrogen levels, one direct consequence of elevated CO_2 that was discussed earlier, also affect decomposition rates. The net effect of all of these variables is difficult to predict. Changes in nutrient availability, brought on by changes in decomposition rates, could have long-term effects on the mixture of species in an ecosystem. Decomposition not only makes nutrients available to other organisms, but it also releases CO_2 to the atmosphere. We leave it as an exercise for you to explain how this response is a possible feedback process between the atmosphere and the biosphere.

Some researchers believe that changes in temperature and rainfall brought on by a changing climate will have their biggest impact by increasing the frequency of fires. The 1988 fires in Yellowstone National Park gave us a graphic demonstration of the possible consequences of dry years. Historical analyses have shown that even small climatic changes during the last few centuries have caused major changes in the occurrence of fires. In fact, researchers can predict fires in Canadian forests by looking at the duration of dry periods. Fires, and other natural disturbances such as pests and pathogens, can change the composition of land communities much faster than migration of species or other processes described above.

Figure 17. Decomposers such as fungi consume dead plant material and return nutrients to the soil. Photo by Carlye Calvin.

Figure 18 shows how fire can affect the reproduction of eucalyptus trees by making more seeds available for germination.

Not surprisingly, it is difficult to predict the result of all of these indirect effects (temperature, rainfall, fire, competition) on the biomes of the world. Some modeling studies have attempted to do so, by using our best guess (from other models) of how temperature, moisture, and CO_2 will change in the future. The studies often predict that temperate forests move northward; boreal forests and subtropical forests decrease in area; and subtropical deserts increase in area (Figure 19). But the models are usually quite limited and often don't include feedbacks and other effects such as migration, competition, or human disturbance of the landscape. Even though modeling remains our best tool for predicting effects on large scales, the results must always be interpreted carefully.

Of course animals (including humans) will also be affected by the changing climate. Plants are at the lower end of the food chain, so animals ultimately depend on them for their survival. Effects are likely to ripple through ecosystems by altering competitive relationships, geographical ranges of species, and species diversity, for example. Ocean biota, too, could show shifts in species ranges and community compositions if warming of ocean water or alterations in water circulation patterns occur.

Changes in seasonal temperature and precipitation patterns could impact the breeding and migration of animals by altering the times when resources are available. Many animal groups, such as birds, whales, fishes, and insects, are highly migratory and take their cues from their physical environment. Some scientists believe that large declines (up to 80%) in populations of several bird species in the eastern United States may be due partly to climate change, which may be upsetting the delicate timing that makes food sources available along migratory paths. Amphibians (frogs, salamanders, etc.) may be sensitive indicators of future climate change because of their critical dependence on moisture and their position in the food chain of the forest.

In the case of humans, the effects of climate change on world food production are of great concern. Water and heat stress can be controlled to some extent by irrigation and the choice of heat-resistant crops, so adjustments will likely be made to minimize the impacts of a changing climate. For example, planners are studying how much heat and moisture corn needs, in order to predict new geographical boundaries for development of this crop. Figure 20 shows a current model's prediction that the North American corn belt will shift northwards, per-

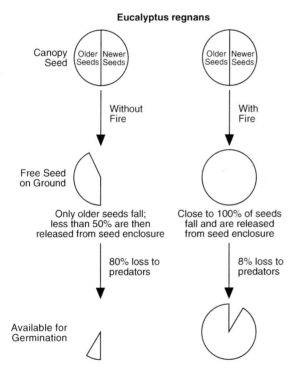

Figure 18. How fire can affect the abundance of the eucalyptus tree. The heat of fire causes more seeds to be released from their capsules than would otherwise be released. Loss of seeds to predators also declines. In this way, fire changes the competitive balance of an ecosystem and enables some species to increase their geographical distribution. Adapted from Chandler, C., P. Cheney, P. Thomas, L. Trabaud, and D. Williams, Fire in Forestry, *Volume 1. Copyright © 1983 by John Wiley & Sons. Reprinted by permission.*

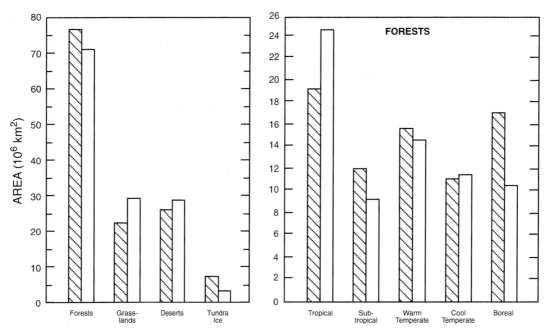

Figure 19. Model estimates of the change in life zone extents for an atmosphere containing twice as much CO_2 as present-day amounts. Present climate is shown in the hatched columns; doubled-CO_2 climate is shown in unshaded columns. Precipitation changes and human alterations of the land surface were not taken into account in the model. From Bolin, B., B.R. Döös, J. Jäger, and R.A. Warrick, eds., SCOPE 29: The Greenhouse Effect, Climatic Change, and Ecosystems, *John Wiley & Sons. Copyright © 1986 by the Scientific Committee on Problems of the Environment (SCOPE). Reprinted by permission.*

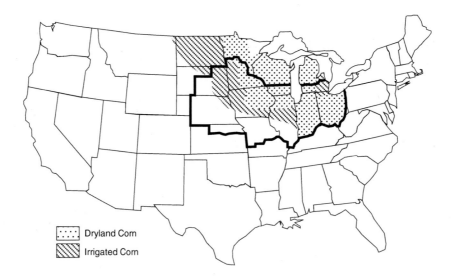

Figure 20. Model estimates of the impact of a 3 °C temperature increase and 8 cm precipitation increase on the geographical extent of the U.S. corn belt. The solid line shows the current location of the corn belt. From Bolin, B., B.R. Döös, J. Jäger, and R.A. Warrick, eds., SCOPE 29: The Greenhouse Effect, Climatic Change, and Ecosystems, *John Wiley & Sons. Copyright © 1986 by the Scientific Committee on Problems of the Environment (SCOPE). Reprinted by permission.*

haps into Canada, but the new borders are highly dependent on precipitation patterns. Winter wheat could replace corn in the southwestern portions of the present corn belt as conditions become warmer and drier. Such adjustments are not automatically possible, however, because soils cannot move with climate zones. Thus, if the optimum climate region for growing a crop shifts, agriculture may not be able to move with it, because the soil may not be suitable. In these cases, another possible strategy is the development of new genetic strains that extend a crop's temperature and moisture tolerances.

Many aspects of human health may be affected by global warming and associated changes in humidity. The temperature of the human body must remain constant within a very narrow range (37°C to 37.5°C). Various thermoregulatory mechanisms ensure maintenance of this core temperature. For example, an increase in body temperature due to exercise is compensated by sweating, which releases heat at the body's surface by evaporation. Capillaries expand, allowing more blood to reach the skin surface, where cooling can occur. Heat is always being generated, even when the body is at rest, from metabolic processes such as respiration and digestion. At high temperature and humidity, the rate at which heat is lost from the body is slowed. This can lead to heat stress and possibly death. People with circulatory problems are particularly susceptible to heat stress. Studies have correlated heart and respiratory diseases, birth defects, infant mortality, and the survival and transmittal of many airborne pathogens (viruses, bacteria, fungi) with temperature, humidity, and other meteorological variables.

We will conclude this section by discussing one last indirect effect of greenhouse gases, which turns out to be quite speculative. Sea levels might rise as surface waters expand and/or glaciers melt because of warmer surface temperatures. By some estimates, sea levels might be 0.5–1 meter higher by the end of the next century. This rate of change would be significantly higher than that experienced over the last 100 years. Though sea-level forecasting involves many uncertainties, it is clear that such a side effect of climate change would have consequences for both humans and coastal ecosystems. A large part of the world's population lives in the coastal zone, within only three meters of sea level. In the United States, the Atlantic and Gulf coasts are low-lying, flat areas that would run a high risk of being inundated. Sandy shorelines are vulnerable to erosion; in fact, a 1-meter rise in sea level could cause some beaches to retreat by as much as 100 meters. In many areas, salt water could intrude into freshwater streams and rivers, threatening drinking water supplies and poisoning croplands. Coastal wetlands, which produce a bounty of species reaped by the worldwide fishing industry, could also be at risk. These marshes, swamps, and mangrove forests are sensitive to water levels, salinity, and wave energy. Marshland could be lost either by erosion or submersion, as sea levels rise. The inland movement of wetlands could be difficult if the terrain is unsuitable or if human alterations of the landscape (cities, dikes, farms) block the migration (see Figure 21).

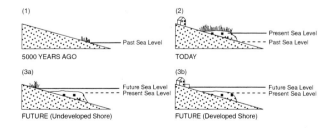

Figure 21. The evolution of marshes as sea level rises. Human development can block the re-establishment of marshes. From White, J.C., ed., Global Climate Change Linkages: Acid Rain, Air Quality, and Stratospheric Ozone. Copyright © 1989, Elsevier Science Publishing, Inc. Copyright © 1993, Chapman & Hall. Reprinted by permission.

Depletion of Stratospheric Ozone

In the early 1970s, scientists first suggested that certain compounds released by human activities could destroy ozone in the stratosphere (the second-lowest atmospheric layer, between approximately 15 and 50 kilometers above the surface of the Earth). It was initially thought that nitrogen oxide gases produced by supersonic aircraft (SSTs) could cause ozone depletion. By the mid 1970s, a new mechanism, halogen pollution, began to receive attention. The compounds of greatest interest, chlorofluorocarbons (CFCs), are used as propellants in aerosol spray cans in some parts of the world, as blowing agents in the manufacture of plastic foams, as cleaning solvents in the electronics industry, and as refrigerants. Research is still continuing to try to determine whether supersonic and subsonic aircraft emissions significantly affect the ozone layer. But strong evidence now shows that chlorofluorocarbons pose a serious threat. Much public attention has been focused on the Antarctic, where scientists have identified an "ozone hole" with as much as 60% of the ozone depleted during the Antarctic springtime. Increasingly, there are also signs of ozone depletion over the more-populated continental regions in the middle latitudes of the globe. Satellite measurements suggest that stratospheric ozone over populated areas of the Northern Hemisphere is now decreasing at the rate of about 1% every two to three years.

The Link Between the Biosphere and the Ozone Layer

How can a change in the amount of stratospheric ozone possibly affect organisms that dwell more than 15 kilometers below it? The depletion of stratospheric ozone could affect the biota indirectly, because ozone plays a fundamental role in shielding the Earth's surface from certain wavelengths of ultraviolet radiation in sunlight. A portion of the UV radiation, known as UV-B and encompassing wavelengths from 280 to 320 nanometers, has harmful biological effects and is only partially blocked by the ozone layer (Figure 22a). Thus, when we discuss the possible biological effects of stratospheric ozone depletion, we are really discussing the potential effects due to UV-B radiation. A 1% reduction of the ozone shield would result in a 1–2% increase in this biologically damaging segment of the radiation spectrum.

Research has shown that UV-B radiation damages the membranes of cells. Much of the UV radiation is taken up by nucleic acids (DNA and RNA [ribonucleic acid]) and proteins in the cells. Since DNA functions as the storehouse of all genetic information, changes in DNA may severely hamper the proper functioning of an individual. RNA is involved in the manufacture of proteins, which play a role in most physiological processes (photosynthesis, respiration, digestion, excretion). In multicellular organisms, the internal tissues and organs are somewhat protected from damaging UV light, because much of the radiation is absorbed by surface layers of cells. The more susceptible organisms are unicellular organisms such as bacteria, single-celled plants and animals, eggs

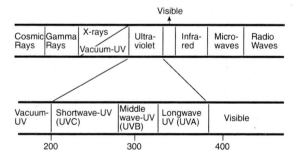

Figure 22a. Invisible ultraviolet light makes up a small part of the electromagnetic spectrum. UV-B, which includes wavelengths from 280 to 320 nanometers, is the most damaging to humans, affecting the skin, eyes, and immune system.

and sperm of aquatic organisms, and individuals who lack protective skin pigments (e.g., albinos). Figure 22b gives what is known as an "action spectrum," which shows how DNA damage increases at the UV wavelengths.

We have some understanding of how UV damages DNA. The structure of DNA consists of two phosphate sugar strands that each have organic bases along their lengths. There are four possible organic bases: thymine, cytosine, adenine, and guanine. The two strands are cross-linked into a double helix structure by pairing the bases along one strand with the bases along another strand, somewhat like a twisted ladder (see Figure 23a). In this cross-linking process, thymine is always paired with adenine and cytosine is always paired with guanine. UV-B radiation can either break the strands or it can disrupt the usual order of the DNA's structure.

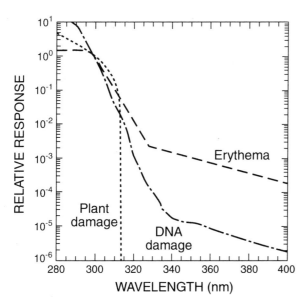
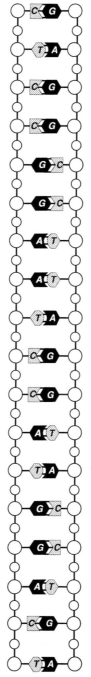
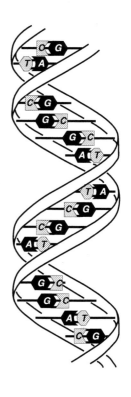

Figure 22b. Action spectra, showing the relative effectiveness of different wavelengths of light in eliciting biological reactions. DNA damage, plant damage, and erythema (sunburning) sensitivities are shown. Adapted from Madronich, S. Implications of recent total atmospheric ozone measurements for biologically active ultraviolet radiation reaching the earth's surface. Geophysical Research Letters 19:1, 37–40. Copyright © 1992, American Geophysical Union. Reprinted by permission.

Figure 23a. The double helix structure of DNA. Note that cytosine (C) always bonds to guanine (G) and that thymine (T) bonds only with adenine (A). From Nelson, G.E., Biological Principles with Human Applications, Copyright © 1980, 1984, 1989, by John Wiley & Sons, Inc. Reprinted by permission.

For example, it has been found to cause incorrect linking between two adjacent thymine units (a "dimer"; see Figure 23b). This altered DNA will then give incorrect instructions to the cell, which can result in some form of damage to the organism.

Organisms have developed some ways to repair such damage. In photoenzymatic repair, the organism uses light and an enzyme to chemically undo the incorrect cross-links. Ironically, the very thing that led to DNA damage, exposure to sunlight, is used by the organism to fix the damage. In another process known as excision repair, the incorrect portion of the DNA sequence is snipped out of the strand and replaced with a resynthesized, correct segment. Both of these repair strategies are widespread. A less common form of repair occurs in bacteria and involves the replication and combination of undamaged strands to form a complete new whole.

Effects of UV-B on Terrestrial Animals

In humans, UV-B radiation plays a beneficial role by converting a form of cholesterol to vitamin D in the skin. However, it is well known that excessive exposure to UV radiation in sunlight can lead to sunburn. Low or moderate exposure produces a reddening of the skin, but higher doses result in the death of more surface cells, which causes pain and blistering. These effects arise mainly from exposure to UV-B. Individuals differ in their susceptibility to sunburn largely due to differences in the pigment content of their skin (Table 2). Pigment-producing cells, called melanocytes, are found in the

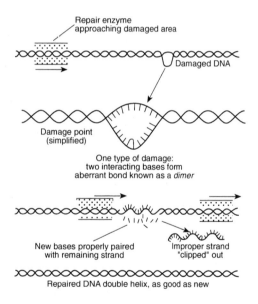

Figure 23b. Acting as a safeguard against the effects of spontaneous damage to DNA, teams of roving repair enzymes travel along the strands, somewhat like railroad inspection crews. The damaged portion is "clipped" out during the repair process. Following this bit of ultrasurgery, bases from the pool are properly paired into a short strand. Then the short strands are fitted into place and bonded. A potentially harmful genetic change (a mutation) is thus eliminated and the DNA is as good as new. From Wallace, R.A., J.L. King, and G.P. Sanders, Biology: The Science of Life, Copyright © 1981 Scott, Foresman and Company. Reprinted by permission of Harper Collins College Publishers.

Table 2 Skin Types and Tanning		
Type	Burning/Tanning Characteristics	Examples
I	Always burns, never tans	Blondes and redheads, blue eyes
II	Usually burns, tans after many hours in the sun	Fair skin, blondes
III	Burns and tans moderately	Most Caucasians
IV	Burns slightly, tans well	Hispanics and Asians
V	Almost never burns, tans darkly	Middle Easterners and Indians
VI	Burns only with very heavy exposure	Blacks

From Gill, P.G., Jr., Skin and sun savvy. Outdoor Life, May 1989. Copyright © 1989 Times Mirror Magazines. Reprinted by permission.

outer skin, or epidermis. They contain the amino acid tyrosine, which is converted to melanin when melanocytes are exposed to UV-B. Melanin, the dark pigment that causes tanning, then acts as a sunscreen that absorbs UV and blocks its transmission to deeper layers of the skin. Fair-skinned people have relatively few melanocytes and thus burn easily. Dark skin, which contains large amounts of melanin, allows much less UV-B radiation to penetrate the skin (Figure 24). Sunscreen products, when applied to the skin, function much like melanin by absorbing the ultraviolet radiation. Physical sunblocks, like zinc or titanium oxide, actually block the UV radiation and scatter it away from the skin. (Some animals and plants also have UV-absorbing pigments that act as protectants. In animals, features such as feathers, scales, and air spaces act to filter or scatter UV radiation.)

Concern about UV-B in humans goes beyond sunburns. Studies show that increased exposure to UV-B radiation promotes the occurrence of squamous-cell and basal-cell carcinomas (skin cancers). Such non-melanoma cancers appear as nodules on the skin surface or as growths that penetrate into the skin. Dark-pigmented skin tumors known as malignant melanomas develop from the proliferation of melanin-containing cells. This is the most dangerous form of skin cancer. The incidence of malignant melanomas has increased substantially during the last 50 years, making them the second most rapidly increasing cancer after lung cancer in terms of mortalities. Ozone depletion is expected to increase the occurrence of fatal and non-fatal skin cancers in the future (Table 3). A sustained 10% decrease in stratospheric ozone would result in an estimated 25% increase in non-melanoma skin cancers worldwide. For malignant melanomas, the relationship to UV-B exposure is more complex and less well understood. Available studies suggest that a 1% decrease in ozone would lead to a 2% increase in malignant melanomas and a 0.3–2% increase in mortality from malignant melanomas.

In general, areas of the skin that are more exposed to the sun have a higher incidence of cancer. Skin cancer also tends to occur more frequently in people who work outdoors, such as police officers, sailors, and farmers. The incidence of skin cancer is also higher among people living at high altitudes or near the equator, where sunlight is more intense. People who receive only occasional exposure to excessive sunlight might actually have a greater risk of cancer than people who routinely work or play outdoors. Light-skinned people have a much higher rate of skin cancer than dark-skinned people.

UV radiation is also detrimental to the human eye. UV-associated damage can come in the form of cataracts or snow blindness. Cataracts are opaque areas that form on the lens of the eye. They can result in blindness as the individual ages. It has been estimated that a 1% decrease in stratospheric ozone will be accompanied by a 0.6 to 0.8% increase in cataracts. Snow blindness results when cells in the outermost lining of the eyeball die and become opaque from overexposure to UV. Fortunately, this situation is not usually permanent. As new cells are regenerated, the eyeball slowly recovers. Curiously, other animals do not seem to suffer from snow blindness as humans do. They do develop squamous-cell skin and eye tumors

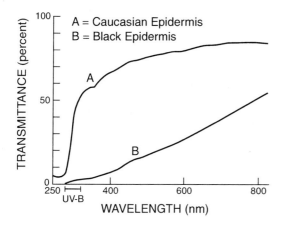

Figure 24. Transmittance of ultraviolet light by human epidermis. From Causes and Effects of Changes in Stratospheric Ozone. *Copyright © 1984 by the National Academy of Sciences. Courtesy of the National Academy Press, Washington, D.C. Reprinted by permission.*

from UV-B exposure. These have been observed in domestic animals such as dogs, cats, goats, sheep, swine, and cattle.

There is one more health concern associated with UV-B, and it is very disquieting. The results of experimental studies with laboratory animals indicate that exposure to UV-B radiation can decrease the effectiveness of the immune system. This change has been related to the development of skin cancers in mice, and there is increasing evidence that such "immunosuppression," as it is called, occurs in humans. This finding is of serious concern because the immune system plays a central role in maintaining and protecting the human body from hazardous chemicals, cancers, and infectious microorganisms. Immunosuppression would tend to make an individual more susceptible to infectious diseases, for example: 1) viral diseases that elicit a rash, such as measles, chicken pox, and herpes; 2) parasitic diseases introduced via the skin, such as malaria; 3) bacterial diseases such as tuberculosis; and 4) fungal diseases. Disturbingly, animal studies indicate that the doses of UV-B needed to induce immunosuppression are lower than those required to produce cancers.

Effects of UV-B on Terrestrial Plants

Scientists have been studying the effects of UV light on terrestrial plants for more than a century. Investigations of the physiological responses of plants have shown that exposure to increased UV-B retards growth, diminishes leaf area, and inhibits photosynthesis. Table 4

Table 3
Summary of Effects of Ultraviolet Radiation on Human Health

Health Effect	Additional Cases*
Non-melanoma skin tumors:	
Additional basal cases	34,035,000
Additional squamous cases	25,848,000
Additional deaths	1,067,000
Melanoma skin tumors:	
Additional cases	422,000
Additional deaths	94,000
Senile cataracts:	
Additional cases	16,869,000
Other skin problems:	(not quantified)
Acute sunburn and thickening of the skin	
Chronic aging of the skin, thinning of the epidermis	
Infectious diseases of the skin (e.g., *Herpes simplex*)	
Other eye disorders:	(not quantified)
Retinal damage	
Corneal tumors	
Acute photokeratitis ("snow blindness")	
Suppression of the human immune system	(not quantified)

*Projection for Americans alive today and born by the year 2075, assuming a 2.5% annual growth rate for CFCs.

From Cogan, D.C., Stones in a Glass House: CFCs and Ozone Depletion. Copyright © 1988, Investor Responsibility Research Center Inc. Reprinted by permission.

summarizes these effects and compares them to the effects of enhanced CO_2. Much of the UV research has focused on crop plants. These differ in their susceptibilities. Squash, cucumbers, and melons are among the most sensitive to UV light. Germinating plants are particularly at risk. Growth is reduced, probably because of the influence of UV on a number of physiological attributes that relate to photosynthesis (for example, chlorophyll concentration and enzyme activity). In some species, there is evidence that UV light inhibits the proper functioning of the stomata. These openings on the surfaces of leaves are responsible for regulating water loss and controlling the exchange of CO_2 and other gases with the atmosphere. Such physiological slowing may be a strategy used by the plant to buy time and conserve resources so that it can carry out the DNA repair processes described on pages 27 and 28.

In the wild, plant populations also differ greatly in their susceptibilities to UV light. UV-resistant populations are found in alpine and tropical habitats, where organisms are exposed to higher amounts of UV. Such plants have developed various mechanisms for reducing the effects of UV. Some have specialized pigments in the outer cell layers of leaves. These act as UV sunscreens, much like melanin in human skin. A few plants defend themselves by producing leaves that are highly reflective. For example the leaves of the silversword plant (Argyroxiphium sandwicense), which grows on the high slopes of Haleakala in Hawaii, reflect

Table 4
Effects of UV-B and CO_2 on Plants

Characteristic	Enhanced UV-B	Enhanced CO_2 (doubling)
Photosynthesis	decreases in many C_3 and C_4 plants	in C_3 plants increase up to 100% but in C_4 plants only a small increase
Stomatal opening	no effects in many plants	decreases in both C_3 and C_4 plants
Water use efficiency	decreases in most plants	increases in both C_3 and C_4 plants
Dry matter production and yield	decreases in many plants	almost a doubling in C_3 plants and only small increase in C_4 plants
Leaf area	decreases in many plants	increases more in C_3 than in C_4 plants
Specif. leaf weight	increases	increases
Drought tolerance	no influence	greater tolerance
Crop maturation	no influence	accelerated
Flowering	some plants faster, others slower	accelerated
Differences within species	response varies among cultivars	response may vary among cultivars
Differences between species	response may vary per community	C_3 plants get more competitive over C_4 plants

From Beukema, J.J., W.J. Wolff, and J.J. W.M. Brouns, eds., Expected Effects of Climatic Change on Marine Coastal Ecosystems. *Copyright © 1990 by Kluwer Academic Publishers. Reprinted by permission.*

about 40% of the light that strikes them. This is the highest reflectivity ever measured for any plant species. UV light may induce some plants to produce UV-screening molecules or to make thicker wax coatings on their leaves. While plants have evolved mechanisms to protect themselves, it is not clear that such defenses would be sufficient to guard against rapid increases in ultraviolet radiation.

There are many remaining uncertainties about the effects of UV radiation on plants, largely because of the experimental difficulties in simulating natural UV exposure. Thus, as in the case of CO_2 research, there are doubts about whether the existing experiments represent what actually occurs in the real world. And what about the effects of UV on the higher levels of biological organization—populations, communities, ecosystems? Predicting the outcome of increased levels of UV on whole communities is difficult because of the many possible interactions among plant species and between plants and animals. Since UV sensitivity varies so widely among different species of plants, it is possible that community composition and structure will be altered. It is certainly difficult to devise experiments to test for effects at these higher levels of biological organization. By default, it seems we are allowing the real world to supply us with an understanding of the consequences.

Effects of UV-B on Marine Life

Considerable attention has been focused on how UV light might affect marine life; after all, the oceans occupy 70% of the Earth's surface. Not only do the oceans' organisms provide us with food, but marine life plays key roles in the production and uptake of important gases in the atmosphere. There has been particular concern about the oceanic plants and animals that dwell in the polar regions, because of the occurrence of the "ozone hole" in the springtime. Biologically damaging UV has been seen to more than double during episodes of ozone depletion in Antarctica.

Most UV light is absorbed in the oceans by dissolved colored pigments and colored particles such as phytoplankton cells. UV penetration of ocean water is usually limited to depths of less than two meters, depending on the amount of these absorbing materials in the water. (Other wavelengths of light penetrate much deeper, as far as 30 meters in clear ocean waters.) There is much concern about the effects of UV-B on phytoplankton, tiny organisms adrift in the upper ocean that produce their own food by photosynthesis. Phytoplankton are at the bottom of the oceanic food chain, which means that they affect virtually every other organism. It is therefore not surprising that a major area of research focuses on the effects of UV light on the photosynthetic processes of phytoplankton. Most work shows that UV light inhibits photosynthesis and reduces growth. The amount of UV that the phytoplankton receives depends on whether or not it is in the upper water layers where UV-B radiation occurs, but it is also highly dependent on vertical mixing of the water column. For a phytoplankton cell that normally lives in surface waters, exposure to UV light will be reduced by turbulence, because the cells will be mixed down to depths that are not reached by the UV light for some portion of the time. But phytoplankton at greater depths will have a greater exposure to UV than they would otherwise.

Studies suggest that not all phytoplankton species will be equally affected by increased exposure to UV light. Thus, scientists hypothesize that increased levels of UV striking the surface of the ocean could lead to changes in the species composition of the phytoplankton community, with those that are resistant increasing in abundance. Such changes could affect zooplankton, the community of organisms that rely directly on phytoplankton as food, and in turn higher levels in the food chain. It is difficult to predict the exact sequence of events, since we still know relatively little about the interactions among species.

Higher marine organisms may also be affected directly by increased exposure to UV.

Other marine plants, such as seagrasses and algae, have been found to be sensitive to UV-B. Animals are also affected. Drifting eggs and larvae of many oceanic animals cannot avoid the upper water layers, and their survival rate declines when exposed to UV-B. Experiments with aquatic animals known as copepods, which are key components of the marine food chain, indicate that both the number of eggs produced and their survival are reduced by exposure to high doses of UV-B. Fish have been found to suffer various kinds of tissue damage and cancers, including skin and gill lesions and carcinomas.

Many marine organisms have evolved mechanisms to reduce the effects of exposure to UV light. Many species that live near the ocean surface (for example, corals) contain UV-absorbing pigments. Alternately, the outer tissues may reflect ultraviolet radiation. (Seagrasses, for example, have this trait.) Other species practice the behavioral mechanism of avoidance, residing at depth during the day and rising to the surface at night. This is known as diel vertical migration. (The burrowing of some shrimp species may be such a defensive response to UV light.) Finally, some marine organisms are able to carry out photoenzymatic repair processes. (Shrimp, anchovies, and seagrasses are examples.)

In summary, the effects of UV-B radiation on a marine ecosystem will depend on:
1) the amount of UV-B reaching the organisms,
2) the UV sensitivities of the various species, and 3) the ability of the organisms to protect themselves by avoidance, production of UV-reflecting or -absorbing molecules, and/or repair of the injury.

Increases in Tropospheric Oxidants

Much attention has been focused on the possible ways in which global change could alter the physical climate at the Earth's surface: temperature, ultraviolet radiation, rainfall, and soil moisture. We have described some of the direct effects of increasing CO_2 on the biota, but the discussion of climate change often neglects other aspects of the chemical climate that living organisms are exposed to. In this section, we will briefly explore how changes in the chemical makeup of the lower atmosphere, or troposphere, could add another dimension to the question of how global climate change might affect life on Earth.

You are probably familiar with some examples of how the chemistry of the lower atmosphere can adversely impact living organisms. Acid rain and urban pollution (smog) are known to cause human health problems and damage to plants and animals. These problems are primarily regional and local, but they do have links to global climate change. For example, the heating of the atmosphere as a result of climate warming will increase the rates of many chemical reactions involved in acid precipitation, and changes in rainfall will alter the way acidic pollutants are deposited on sensitive ecosystems. Despite these links to global issues, we can be thankful that the impacts of acid rain and urban pollution remain on local and regional levels. For this reason, we will not discuss these problems in this module. We will note only that they are serious problems, deserving of study by scientists and action on the part of policy makers.

We have already discussed the fact that greenhouse gases such as CO_2, methane, nitrous oxide, and chlorofluorocarbons are building up in the atmosphere. This is not disputed in the scientific community, as the experimental evidence is overwhelming. There is, however, controversy among atmospheric chemists over how the composition of the lower atmosphere might be changing on a global basis. It is thought that tropospheric oxidants—gases such as ozone and hydrogen peroxide—might be increasing. Oxidants are reactive molecules that play a crucial role in the chemistry of the atmosphere, but they have another potential importance: they could have direct effects on living organisms. These oxidants react with many molecules in the tissues of plants and animals.

What is the evidence that oxidants are increasing in the lower atmosphere? Most of the work has focused on ozone, a molecule with which you are now somewhat familiar. Note that we are making a distinction between stratospheric ozone (which acts as a filter to absorb harmful UV-B radiation) and tropospheric ozone (which comes into direct contact with the biota). Scientists have been attempting to make measurements of tropospheric ozone since the 19th century. The earliest measurements may have had some problems, but modern scientists have attempted to correct the data to account for the difficulties. The record now suggests that the level of background ozone has doubled over the last 50–100 years (see Figure 25). Background ozone concentrations are not part of a local or short-lived pollution episode. They represent the clean troposphere—a sort of "best case" for the globe.

In the case of hydrogen peroxide (H_2O_2), there are no historical measurements for us to refer to. This molecule turns out to be very difficult to measure, and in fact it has only been measured reliably since the mid-1980s. But scientists have analyzed air bubbles trapped in polar ice cores (Figure 26). By drilling cores deep into the ice, one is essentially traveling in time to earlier climates and earlier atmospheres. The ice core record for H_2O_2 suggests that this gas has increased in abundance over the last few decades.

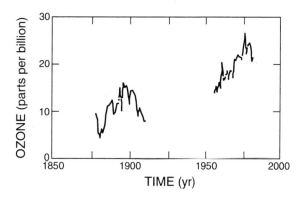

Figure 25. Annual mean ozone concentrations at two rural sites in Europe: Montsouris (1876–1910) and Arkona (1956–83). From Volz, A., and D. Kley, Nature *332: 240. Copyright © 1988, McMillan Magazines Limited. Reprinted by permission.*

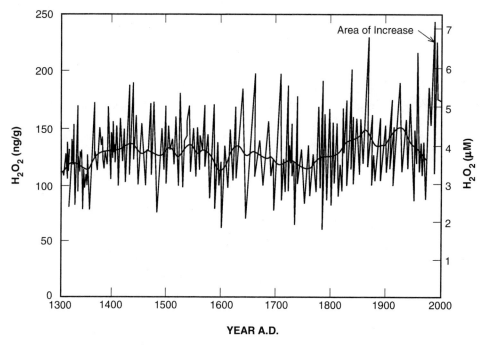

Figure 26. Recent increase in hydrogen peroxide (H_2O_2) dissolved in an ice core from Greenland. From Thompson, A. The oxidizing capacity of the earth's atmosphere: Probable past and future changes. Science *256: 1157. Copyright © 1992, AAAS. Reprinted by permission.*

Figure 27 shows just a few of the key reactions that lead to the production of these oxidants. The detail isn't important, but you can see that the basic chemistry starts with materials such as carbon monoxide (CO) and volatile organic compounds including methane (CH_4), which undergo a series of chemical reactions resulting in O_3 and H_2O_2 products. Nitrogen oxides (NO and NO_2) are also key players in these reactions. Because of human activities, CO, CH_4 and other volatile organic compounds, and nitrogen oxides are increasing in the atmosphere. Computer models of atmospheric chemistry tell us that these increases combine with other features of climate change (such as increased tropospheric water vapor and stratospheric ozone depletion) to cause an increase in tropospheric O_3 and H_2O_2. Figure 28 shows the results of a chemical model that predicts large increases in these oxidants in the coming decades.

What are some of the possible biological consequences of a changing oxidant climate in the troposphere? The effects of ozone have received the most scientific attention. For several decades, researchers have studied the effects of O_3 on plants. It is often found that ozone reduces photosynthesis, decreases growth, causes deterioration in cell membranes, and discolors or kills leaf tissue (see Figure 29). In severe exposure, individual plants can suffer the ultimate penalty of death. Figure 30 shows a schematic of various ways in which ozone can affect leaf tissue.

Ozone effects depend on many factors, such as genetics, plant and/or leaf age, environmental conditions during exposure, ozone concentration, and the presence of other pollutant gases. Some species (such as some varieties of tobacco) are far more sensitive than others, a phenomenon that can't be predicted. Even within the same species, some individuals are more tolerant than others. The best-documented case linking vegetation damage directly to ozone is in the San Bernadino Mountains east of Los Angeles, California. There, researchers have been studying the damaging effects of ozone on ponderosa pine and other forest trees. It was found that ozone does not kill the trees outright; rather, it weakens them to the point where other factors such as insects or root rot

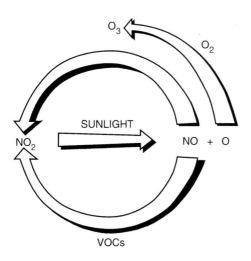

Figure 27a. Simplified scheme of reactions that lead to ozone (O_3) formation and buildup. Sunlight and oxygen (O_2) convert nitrogen dioxide (NO_2) to ozone. Volatile organic compounds (VOCs) reduce ozone consumption by competing with ozone for nitric oxide (NO). From ROPIS Newsletter summer 1991. Copyright © 1991, Electric Power Research Institute, Inc. Original figure designed by R.A. Goldstein and A. Hansen, EPRI. Reprinted by permission.

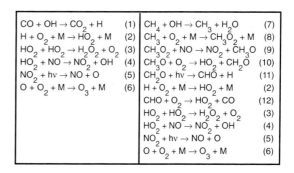

Figure 27b. Simplified chemical scheme showing how the oxidants O_3 and H_2O_2 are produced from the photochemical oxidation of methane (CH_4) and carbon monoxide (CO). The symbol "hn" represents ultraviolet light, and the symbol "M" represents any molecule in the atmosphere.

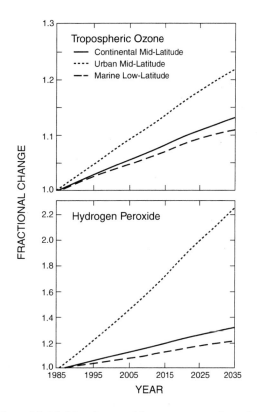

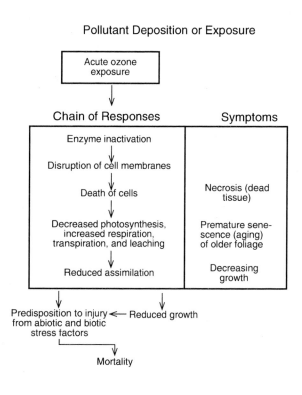

Figure 28. Model estimates of future concentrations of tropospheric ozone and hydrogen peroxide at various latitudes, as compared to the 1985 values. At urban mid-latitudes, H_2O_2 is projected to be over twice as concentrated in the year 2035. (Courtesy of Anne Thompson, NASA Goddard Space Flight Center).

Figure 30. Hypothesized pathway involving direct ozone damage to foliar tissue. From World Resources 1986, *New York: Basic Books. Copyright © 1986 by the World Resources Institute and the International Institute for Environmental Development. Reprinted by permission.*

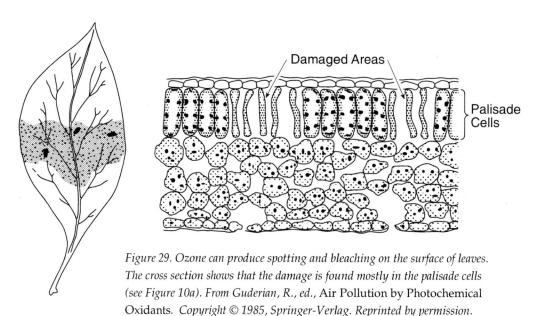

Figure 29. Ozone can produce spotting and bleaching on the surface of leaves. The cross section shows that the damage is found mostly in the palisade cells (see Figure 10a). From Guderian, R., ed., Air Pollution by Photochemical Oxidants. *Copyright © 1985, Springer-Verlag. Reprinted by permission.*

lead to mortality. Figure 31 shows some of the ways that O_3 damage to trees impacts other components of the forest ecosystem.

In addition to natural vegetation, agricultural crops are also damaged by ozone. Ozone damage to crops in the United States alone costs an estimated one to two billion dollars per year.

Most of the research on ozone effects on plants has used concentrations of ozone well above the background levels of 20–40 parts per billion (ppb). Thus, the gradual rise in background concentrations may not impact plants for some time. But some researchers point out that the gap is rapidly narrowing between the tolerance levels of some plants and the background O_3 concentrations that are now being measured. In fact, this gap may be the smallest for any atmospheric pollutant. Thus, the future may find us dealing with our first truly global air pollution problem.

In humans, ozone irritates the nose and throat and impairs lung function in both healthy people and those with asthma and other chronic respiratory problems. Animal studies suggest that prolonged O_3 exposure could damage cells and accelerate the aging of lung tissue. However, most research has been conducted using O_3 concentrations well above the present and predicted future background concentrations. Thus, human health effects will likely remain a problem only in localized areas that experience episodes of high ozone.

As for the effects of the other oxidant we mentioned, hydrogen peroxide, we are virtually without a clue. This molecule is very difficult to work with and measure, and it is only recently that some research has been attempted. The few existing experiments show that the hydrogen peroxide gas is so reactive that it may never enter plant leaves; it is destroyed on the leaf surfaces before it even enters the stomata. Thus, this gas has not yet been shown to affect internal processes such as photosynthesis and respiration. An unanswered research question concerns what external effects H_2O_2 might have as it is deposited on the leaf. It will be important to study this topic, since Figure 28 suggests that H_2O_2 may be increasing at higher rates than ozone.

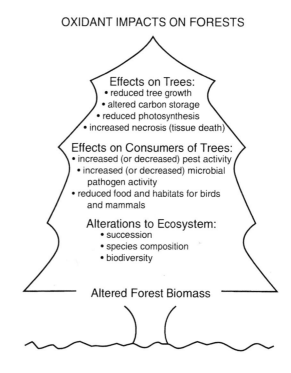

Figure 31. Cascading effects of oxidants on trees and other aspects of the entire forest ecosystem. Evidence is stronger for some effects than others; some evidence is contradictory or dependent upon the particular species involved.

III
Uncertainties and Challenges for Future Research

We have discussed many of the possible effects of global climate change on the biota, but we have really only scratched the surface of this topic. Entire books have been written on several of the individual interactions we have mentioned, so our review is not all-inclusive. It has been our hope to give you a feeling for the wide range of possible consequences that are of concern to all of us.

While we have concentrated on how global change might impact the biosphere, it is important to note that the biosphere is not just a passive receiver of these impacts. Instead, it is an active player in the total global change picture. For example, several of the greenhouse gases have biological sources: CO_2, methane, nitrous oxide. Another example of the close coupling of the biosphere and the chemical climate is that some plants emit hydrocarbon gases that contribute to the formation of tropospheric ozone. Oceanic phytoplankton emit sulfur gases that form aerosols in the atmosphere, which, in turn, influence cloud formation and precipitation.

Clearly, the interplay of climate and life is complicated and there are many possible feedbacks to consider. For example, climate warming could increase the rate of decomposition, which would release more CO_2 into the atmosphere and contribute to further warming (a "positive feedback"). One of the current hot topics in scientific research involves the biosphere as an active participant in the global climate. It seems that the rate at which CO_2 is increasing in the atmosphere is slower than what scientists would expect based on our knowledge of how much we are adding to the atmosphere through fossil fuel burning and biomass burning (deforestation). Where is the missing CO_2 going, if not to the atmosphere? Some scientists believe that the answer lies with the boreal forests of the Northern Hemisphere, where ecosystem responses to global change may have altered the way CO_2 is cycled. Much research is now being conducted in an effort to identify this missing carbon storage component (the so-called "missing carbon sink").

Contemplation of the biological consequences of climate change is an unsettling exercise. There are many uncertainties in each of the possible biotic responses we have described, adding stress to the political, economic, and psychological aspects of our global dilemma. In this short space, we have not even explored what is, perhaps, the most challenging aspect of climate change: the fact that all of the changes will be occurring simultaneously. Thus, we must consider that CO_2 increases, temperature changes, rainfall changes, UV-B increases, and increases in oxidants will be superimposed all at once on the biota. What will happen to individuals or single species faced with multiple changes in the physical and chemical climate? And what implications will the multitude of changes have for the intricate intertwined relationships of organisms at the ecosystem level? The possible number of interactions is staggering. Some may counteract each other; for example, the increase in plant growth from CO_2 fertilization and the decrease in plant growth from reduced annual precipitation. Others could reinforce each other; increased UV-B radiation and increased tropospheric oxidants might both tend to reduce photosynthesis in plants, for example. Or, in combination, multiple stresses might act in entirely new ways that we can't predict from our study of their individual effects.

Not surprisingly, with so many uncertainties about the effects of the individual climate factors covered in this module, the research into the effects of simultaneous multiple factors is in its infancy. There is much to be done. It will be the challenge of your generation to advance our scientific understanding, and to embrace a lifestyle that ensures a sustainable future for all living organisms.

Copyright © 1993 by David Kolosta, Houston Post. Reprinted by permission.

Problems and Discussion Questions

1. If the world's vegetation is fertilized significantly by atmospheric CO_2, some scientists believe that the amount of CO_2 consumed during the growing season will increase. If this occurred, how would you expect the sawtooth structure of Figure 8 to be changed? Do you see any evidence for this effect in the data of Figure 8? (Hint: Use a ruler.) What other real-world factors could complicate the plants' response to increased CO_2?

2. Extensive tree planting has been proposed as one means of absorbing excess atmospheric CO_2 and slowing global warming. The plot below shows the carbon uptake ability of a tree as a function of time, from seedling to maturity. Comment on the effectiveness of the tree-planting strategy in the: (a) short term; and (b) long term.

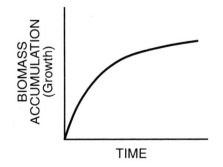

3. Which plant shown below would not be expected to show a big increase in photosynthesis when CO_2 is increased from present-day values (350 parts per million, ppm) to 700 ppm? Is plant A likely to be a C_3 or a C_4 plant?

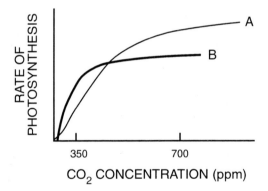

4. The figure below shows the geographical ranges of two species, A and B. Discuss how climate change affects species diversity at the point shown, if species A responds to climate change by migrating northward but species B cannot migrate.

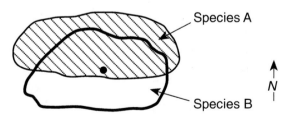

5. One aspect of a species' response to warming is that in mountainous areas, a species may shift altitudinally as well as latitudinally. Generally, a short climb in altitude corresponds to a large shift in latitude. For example, a 3° C cooling can be achieved by either a 500-meter climb in elevation or a 250-kilometer shift in latitude. The figure below shows the present altitudinal distribution of three hypothetical species, A, B, and C.

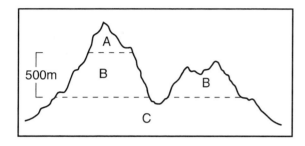

   a. Draw the new distribution of species in response to a 3° C climate warming, assuming the shifts mentioned above.

   b. How is the size of each population likely to be affected as a result of the shifts?

   c. What factors could prevent the species from accomplishing such a shift in response to a human-induced 3° C climate warming? (Assume that all three of the species are plants. Then, answer as if the three species were animals.)

   d. Suppose that Species B is sensitive to UV-B radiation. Describe how this could affect the migration of this species in response to climate warming. What other global change could complicate the adjustment of Species B to climate warming? Why?

6. Why will increased levels of carbon dioxide in the atmosphere not necessarily lead to increased levels of photosynthesis?

7. How does the development of heat-tolerant strains of agricultural crops by scientists mirror the process by which species in natural ecosystems will respond to global warming?

8. Discuss some general types of organisms that would be sensitive to any changes in wind patterns or velocities brought on by global change.

9. Most research into the effects of physical or chemical climate change on trees is carried out on seedlings in greenhouse or laboratory settings rather than on mature trees in nature.

   a. What are the advantages of using seedlings in a managed setting?

   b. Describe some ways the results of these experiments may not be a good indication of what will happen in the "real world."

10. The metabolic processes of plants and animals are temperature-dependent. This relationship is often expressed using a temperature coefficient called Q_{10}, a factor that shows how the rate of a process increases with a 10° C increase in temperature. The formula for Q_{10} is:

$$Q_{10} = (R2/R1)^{10/(t2-t1)}$$

   where R1 is the rate at temperature t1 and R2 is the rate at the higher temperature t2. The figure below shows how the rate of photosynthesis of two plant species depends on temperature. Using the figure,

   a. Estimate the Q_{10} for a base temperature (t1) of 20° C for Species A.

   b. Repeat for Species B.

c. Discuss how a warming of the average summer temperature from 20° C to 23° C might affect the competitive relationship of these two species.

d. Photosynthesis is a process that uses several different enzymes. Explain why, in the curves below, photosynthesis doesn't increase indefinitely. (That is, why do the curves turn over?)

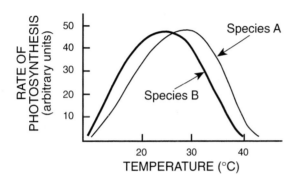

11. Discuss differences in how deforestation affects global warming and the carbon cycle if (a) the trees are burned; or (b) the trees are used to build houses.

12. Use Table 4 (page 31) to explain why the simultaneous occurrence of increased CO_2 and stratospheric ozone depletion might have counteracting effects on plants. What other aspects of global physical/climate change could complicate this simple analysis?

13. It is thought that the buildup of greenhouse gases might lead to a warming of the troposphere but a cooling of the stratosphere. In the stratosphere, cooler temperatures would tend to slow down the chemical reactions that destroy ozone. Discuss the indirect effects that stratospheric cooling could have on the biosphere.

14. What adjustments might occur in the carbon cycle to keep the amount of carbon dioxide in the atmosphere constant as more CO_2 is released from the burning of fossil fuels? Use Figure 7.

15. Scientists believe that an increase in ultraviolet radiation would cause more ozone to be produced in the lower region of the atmosphere (the troposphere). This is because ultraviolet wavelengths of light have sufficient energy to break molecules apart; they "kick-start" some of the reactions that lead to ozone production (see Figure 27).

    a. Explain how this is an example of the interrelationship between two global change processes.

    b. What stresses would be simultaneously imposed on the biota by this relationship?

16. What are your personal views about the issues raised by Michael Oppenheimer in the article "Vulnerable Ecosystems" (Appendix 1)?

Note to Instructor: Suggested Class Activity (Class Debate)

As a class, watch the video *The Greening of the Planet Earth*. Assign working groups to research various scientific issues raised in the video, and then stage a classroom scientific debate. The video is available at no charge from Western Fuels Association, Magruder Building, 1625 M St. N.W., Washington, D.C. 20036 (202-463-6580).

Appendix

*Vulnerable Ecosystems**
Michael Oppenheimer
Environmental Defense Fund

Abstract

Climatic change due to anticipated greenhouse warming will affect ecosystems through changes in temperature, precipitation, soil moisture, storm frequency and intensity, and sea level. For slowly migrating species, including trees in mid-latitude forests, successful adaptation to expected rates of temperature change may not be possible. Even for gradual rates of temperature change, substantial shifts in forest type and biomass will occur according to analyses of anticipated life zone changes. Human barriers to migration of plants and animals will further inhibit successful adaptation, particularly for species currently restricted to "island refuges" such as national parks. The continuous nature of projected climatic warming will inhibit attempts to ameliorate these difficulties. Decreased soil moisture in mid-continent areas could reduce habitats for species dependent on wetlands.

Sea level rise will remake coastal ecosystems. Loss of wetlands and barrier beaches in developed areas will occur due to inhibition of inland migration by settlements and other land uses. Reduced river flow due to upstream diversions will combine with higher sea level to disrupt the coastal environment completely in heavily populated delta regions, particularly at low latitudes. Increased tropical storm frequency or intensity would also be highly problematic in the humid tropics, while semi-arid regions would be vulnerable to changes that either decrease soil moisture or increase the variability of precipitation.

The stresses on ecosystems due to climatic change cannot be separated from other problems already affecting them. Regional air pollution has significantly disturbed mid-latitude forests. Human population pressures are destroying forests in the humid and semi-arid tropics. Excessive nutrient supply is leading to the gradual eutrophication of coastal seas and estuaries. Climatic change will exacerbate these stresses continuously at an increasing rate. Such interactive effects may already be observable, including synergistic disturbance of ecosystems. One net consequence of these stresses may be a decrease in terrestrial biomass accompanied by an increase in biomass, sedimentation, and eutrophication in coastal ocean ecosystems.

Introduction

I would like to thank the organizers of this conference for inviting me to speak about the consequences of global warming for natural ecosystems because it is natural ecosystems that will bear the full brunt of climatic change. It has

---

*From *Ozone Depletion: Health and Environmental Consequences,* edited by R. Russell Jones and T. Wigley. Copyright © 1989 by John Wiley & Sons, Ltd. Reprinted by permission.

been argued by many experts that agriculture may well adjust to climatic change without disruption of global food supplies. It may be argued that human societies as a whole will adapt, particularly if the change is not too rapid, although not without considerable pain. However, it is difficult to support the notion that rapid global warming will entail anything less than a disaster for may natural systems. It is certainly plausible to argue that climatic change will remake the face of the earth.

I want to discuss some of these consequences in this lecture. This discussion is largely based on the results of the Villach 1987 conference, Developing Policies for Responding to Climatic Change[1]. The papers from this conference will be published as a special issue of *Climatic Change*. However, it should be understood from the outset that, if our quantitative understanding of the potential climate changes is limited, our knowledge of ecological effects is primitive. Given the uncertainty, this discussion is really about vulnerability, as the title indicates, rather than quantitative estimates of effects. Nevertheless, there are three general principles of global warming from which the notion of disastrous change proceeds, even while the specifics remain uncertain. They are as follows:

(1) Climate change will be continuous as long as greenhouse gases are emitted in quantities anywhere near current emissions.

(2) Based on current predictions, the sustained rate of global warming may well exceed both recent and more remote historical rates by a large margin.

(3) Climate change is far from the only stress occurring in a natural world that is already under siege from air and water pollution, deforestation, soil erosion, and other human pressures.

I shall return to these points later, while I focus on only two types of ecosystem: mid-latitude forests of the type that cover much of Europe and the USA, and coastal ecosystems such as wetlands and barrier beaches. But remember that the effects of the warming will be global.

Let us set the stage for this discussion by examining once again a projection of future global mean temperature. The recent Villach-Bellagio analysis projected that the global mean temperature would most likely increase at a rate of about 0.3°C/decade using a business-as-usual emissions scenario, with the warming rate perhaps twice as fast at high latitudes. For comparison, except for short periods, the historical rate of change of global mean temperature may not have exceeded about 0.1°C/decade either in recent times or during the last glacial retreat.

In other words, in the future, ecosystems may be required to respond much more rapidly than in the past to climatic change.

Consequences of Global Warming

Consequences of global warming are imposed through *local* changes in temperature, precipitation, and storm patterns. For instance, several models[2] have been used to project local changes in summer season soil moisture, which governs runoff and water availability for plants and animals. In at least one model, summer soil moisture is projected to decrease over much of the globe, particularly the mid-continent areas, over the next 100 years in the business-as-usual scenario. However, there remains controversy over the extent of this drying, with different models giving substantially different results, or even similar results for different reasons.

The clearest way to understand the consequences of these changes for ecosystems is to consider where things live now and where they may possibly find hospitable climate conditions in the future. One simple way to do this is by examining Holdridge life zones[3] derived from temperature and precipitation. Within 100 years, the warming could eliminate the Arctic ecosystem from Alaska and move other zones by about 500 km to the north, according to this analysis. Species restricted to island refuges, such as national parks or mountaintops, may simply have nowhere to shift to and may disappear. Forests along the southern margins of Holdridge zones will shrink, while those at the northern end could expand.

The situation is complicated, and probably further exacerbated by the following:

(1) Ecosystems don't move as a piece; they scatter. So ecosystems won't shift; they will change.

(2) Soils aren't always appropriate at the latitudes that species would be pressed to move toward.

The combined effect of these two factors is projected to bring biomass declines of one third to one half in U.S. southern pine forests, for instance[4], over the next century.

(3) Trees can migrate only slowly. Trees moved at 20–200 km/century during the glacial retreat, a rate that may indicate the limit of their ability to move. Also, human barriers now intervene. Compare that rate with the 400 km/century derived from the Villach-Bellagio scenario. In other words, many tree species may simply disappear because they cannot compete with the rate of climate zone shift.

There are important uncertainties that require exploration. Will increased carbon dioxide "fertilize" perennials as it does some crops? Or will weeds develop a selective advantage? Can human intervention through seed-spreading sustain trees or will silviculture itself fail in the face of rapid warming and zone-shifting?

How may future terrestrial systems actually look? Wetland wildlife habitats will dry up where summer soil moisture is decreased. Wildlife is particularly vulnerable to consecutive dry years, and increased dryness will increase the probability of successive dry years. Forests are already declining in Europe and North America, in part due to air pollution. What will happen when a warming stress is added? Synergistic effects can be expected to accelerate the current decline, and the forest fires of this summer in the USA [1988] remind us that forests do not die quietly, particularly in a dry world.

Now let's examine another type of ecosystem: that of the coastal environment. Here we have the wetlands and estuaries which are the homes for migrating birds and the nurseries for half the fish caught for food. Here we have beaches. Half of humans now live near coasts, and this fraction is increasing. In Asia, they often live in low-lying, heavily cultivated deltaic regions, particularly vulnerable to flooding. In the coastal zone, fresh drinking water and salt water lie close together. The coastal zone is also the receiver of all upstream changes, such as those that might occur in nutrient and sediment flows due to forest decline. This coastal environment will be exposed to the combined effects of a number of stresses; higher temperature and sea level, higher carbon dioxide concentrations, and higher nutrient flows. Coastal ecosystems may be, in short, remade and destroyed. Sea level may rise much faster than the recent historical rate due to expansion of ocean water and melting of land ice. Societal adaptation by movement or protection of infrastructure and beaches will be costly. In deltaic areas such as Bangladesh, resources for adaptation are limited, so abandonment and increasing loss of life can be expected.

Once again, it is the natural ecosystems that simply will not adjust successfully. Wetland migration is unlikely. A one third to two thirds loss in U.S. northeast coast wetlands has been projected[5]. Areas like the Baltic and North Sea, already under stress due to a nutrient excess, will be further damaged by algal blooms encouraged by increased temperature, carbon dioxide, and nutrients. At the very least, species composition changes will occur. Outside the mid-latitudes there are other vulnerabilities. Low-latitude vegetation is particularly vulnerable to climate variability where deforestation and soil erosion are already a problem. Climate feedbacks involving ice and tundra at high-latitude regions could accelerate the warming and create additional stresses for ecosystems.

Is there any good news? Some ecosystems might benefit (if any change is beneficial) as measured by biomass increase. The northern boreal forests may provide an example, if they can move fast enough and if the warming stops at some point.

Conclusion: The Need for Emissions Abatement

Let me summarize by putting these changes in context. Ecological change has occurred before. Recovery from the last glacial age is still occurring for some species. Cyclic stresses like fires are important and natural.

However, global warming will probably occur faster and be more sustained than previous climate changes and will not stop to allow stabilization or adjustment. We are moving into a world of continuous change in which slow-moving species will drop by the wayside and entire ecosystems will disappear. Uncertainties are very large, but because consequences lag far behind the emissions responsible we cannot wait to observe signs of ecological decline and then pull back. If we don't like the consequences of warming after the fact, our recourse is limited since the changes are, in a practical sense, irreversible on a human time scale.

The biosphere is also a storehouse of potential greenhouse gases—carbon dioxide, nitrous oxide, and methane. The warming and associated disturbance of the biosphere could well accelerate the growth of greenhouse gases in the atmosphere and, thus, cause further increases in the rate of warming.

Finally, let me finish on an optimistic note. These changes are not inevitable and the future remains, by and large, a matter of choice.

Two sorts of emissions "futures" are imaginable: one, business as usual; another, a course of emissions, determined by substitution of efficiency measures and renewable energy for current use patterns, that would slow global warming to a pace to which ecosystems might adapt and which may ultimately stabilize the climate. Achieving that second course is a daunting challenge, but I hope everyone will keep in mind that human beings and societies are ultimately dependent on the biosphere. And as ecosystems go, so ultimately goes the human race.

References

1. J. Jager, *Developing Policies for Responding To Climatic Change*, Proceedings of conferences at Villach, Austria, 28 September to 3 October 1987 and Bellagio, Italy, 9–13 November 1987, World Meteorological Organization and United Nations Environment Programme, 1–53 (1988).

2. J.F.B. Mitchell, The greenhouse effect and climate change. *Rev. Geophys.* 27, 115–139 (1989).

3. W.R. Emanuel, H.H. Shugart, and M.P. Stevenson, Climatic change and the broadscale distribution of terrestrial ecosystem complexes. *Climatic Change 7*, 29–43 (1985).

4. S.S. Batie and H.H. Shugart, The consequences of climate change on environmental resources: an economic and biological assessment. Prepared for workshop on Controlling and Adapting to Greenhouse Warming, Washington, D.C. (June 1988).

5. R.A. Park, M.S. Trehan, P.W. Mausel, et al., *The Effects of Sea Level Rise on U.S. Coastal Waters*, Holcomb Research Institute, Butler University, Indianapolis, Indiana, pp. 1–59 (1988).

GLOSSARY

abiotic environment—The nonliving physical and chemical environment that exerts an influence on living organisms; for example, soil, water, air.

acclimation—The process by which an individual organism is able to adjust physiologically to a change in immediate environmental conditions.

aerobic—Requiring molecular oxygen (O_2) for life or activity. Taking place in the presence of O_2.

algae—Primitive water-dwelling plants with a one-celled or simple multicellular structure. Algae lack roots, stems, or leaves but usually contain chlorophyll. They include kelps and other seaweeds, and phytoplankton such as the diatoms.

anaerobic—Living or active in the absence of oxygen.

basal cell carcinoma—A relatively common type of skin cancer that can result from exposure to sunlight. This cancer arises in the basal cell layer of the epidermis, where it spreads but rarely metastasizes to other organs.

biome—Major type of land community of organisms, usually identified in terms of characteristic vegetation; for example, tropical rain forest or desert.

biosphere—All areas on Earth where living organisms are found. Includes oceans, land, and part of the atmosphere.

biotic—Having to do with living organisms.

C_3 plants—Plants that make photosynthetic intermediate organic substances containing three carbon atoms. These plants cannot concentrate CO_2 near the enzymes of photosynthesis, and they tend to lose energy through a process called photorespiration. Examples: wheat, algae, most tree species.

C_4 plants—Plants that make photosynthetic intermediate organic substances containing four carbon atoms. These plants are able to concentrate CO_2 near the enzymes of photosynthesis, and they tend not to lose much energy through photorespiration. Examples: corn, grasses, tropical plants.

Calvin-Benson cycle—A series of enzyme-catalyzed reactions that occur within C_3 plants during photosynthesis. The reactions convert CO_2 into more complex, energy-storing molecules.

carbon cycle—The complex array of chemical and physical processes by which carbon flows through the biotic and abiotic environment.

carcinoma—A cancer of the surface tissues of an organ or organism. Carcinomas include basal and squamous cell tumors, among others, reflecting the type of cell from which the cancer originates.

catalyst—A substance that alters the rate of a chemical reaction, while itself remaining chemically unchanged by the reaction. A biological catalyst is called an enzyme.

chlorophyll—The organic green pigment molecule, occurring in plants, that captures light energy and channels it into the photosynthetic process.

community—All of the organisms living in a particular habitat.

copepod—A small marine or freshwater crustacean. Copepods are an important component of the zooplankton community in most parts of the world's oceans and lakes. They are often referred to as the insects of the sea.

CO_2 fertilization effect—The increase in plant photosynthesis and/or growth that may occur in the presence of increased levels of CO_2 in the atmosphere.

decomposition—The decay of dead organic material. Decomposition is carried out by microorganisms and other decomposers that break down complex molecules and release the nutrients and the carbon (as carbon dioxide or methane).

diel vertical migration—The process carried out by many zooplankton and fish in oceans and lakes, whereby individuals migrate between subsurface layers and the surface. Animals are typically at depth during the day and migrate to the surface at night to feed. Such migration can be used to avoid UV exposure.

digestion—The chemical process whereby life forms break down food molecules into smaller, simpler molecules.

DNA—Deoxyribonucleic acid, the genetic material of almost all living organisms. DNA structure is a double helix of two phosphate-sugar strands cross linked by pairs of organic bases.

ecosystem—The unit of ecological organization that includes the entire community of organisms plus the physical environment in which it occurs.

enzyme—A specialized protein used by living organisms to affect the rates of various physiological processes.

epidermis—The outermost layer of the skin.

excretion—The process by which waste products or undigested food products are eliminated from living organisms.

feedback—The return of a portion of the output of any process or system to the input. The return may either add to the initial input (positive feedback) or subtract from the initial input (negative feedback).

genotype—The inherited constitution of an organism, which determines its characteristics and development.

herbivore—An animal that eats plants or parts of plants.

homeothermic—Regulating and maintaining a constant body temperature through internal mechanisms; warm-blooded.

immune system—That system of cells and molecules responsible for protection of the body of an organism from invasion by foreign substances, organisms, and cells.

immunosuppression—The inhibition or hindrance of the natural immune system of an organism.

malignant melanoma—An often-fatal form of skin cancer that is the result of transformations of the melanocytes in the skin. Four different types have been identified: superficial spreading melanoma, nodular melanoma, Hutchinson's melanotic freckle, and unclassified melanoma.

melanin—A dark skin pigment, manufactured by melanocytes, that absorbs ultraviolet radiation and protects deeper skin layers from exposure.

melanocytes—Specialized cells in the human skin that manufacture melanin when exposed to UV light.

nanometer—A unit of length, equal to one-billionth of a meter (10^{-9} meters).

net primary production (NPP)—Organic matter accumulation in plants, calculated as the difference between photosynthesis and respiration.

non-melanoma skin cancer—A cancer that affects the keratin-producing cells in the skin, including basal cell and squamous cell carcinomas. This kind of cancer has a higher cure rate than melanoma.

organic substances—Chain-like or ring-like carbon-containing compounds that also contain hydrogen and can contain oxygen, nitrogen, sulfur, and phosphorus.

oxidant—A reactive chemical compound that tends to lose electrons when it interacts with other molecules. In the atmosphere, important oxidants are ozone (O_3), hydrogen peroxide (H_2O_2), hydroperoxyl radical (HO_2), and hydroxyl radical (OH).

PAR—Photosynthetically active radiation. Visible light in the wavelengths from 400 to 700 nanometers (blue to red light), used in the process of plant photosynthesis.

part per million (ppm)—A unit that expresses the concentration of a chemical substance. A 1 ppm concentration of molecule X means that one molecule out of a million is molecule X. Related units are part per billion (ppb) and part per trillion (ppt).

pH—A numerical scale that indicates the acidity of a substance. Neutrality is indicated by pH 7; higher pH values are basic, and lower pH values are acidic.

photochemical reaction—Any chemical reaction that requires light (often ultraviolet light).

photorespiration—Plant respiration that occurs in the presence of light. Photorespiration wastes energy that could otherwise be used to make photosynthetic products. It is common in C_3 plants in bright sunlight on hot days.

photosynthesis—The process by which light, carbon dioxide, and water are used by plants to make carbohydrate products, essentially converting light into chemical energy; also called primary production.

phytoplankton—Tiny photosynthetic plants that live in the water column of oceans and lakes. Phytoplankton form the base of the aquatic food chain.

poikilothermic—Unable to regulate body temperature by internal mechanisms; cold-blooded.

population—All members of a particular species in a given area at one time.

protein—Any of a group of organic molecules that occur in all living matter and are essential for the growth and repair of tissue. Proteins contain amino acids as their fundamental building blocks. All proteins contain carbon, hydrogen, nitrogen, and oxygen; nearly all contain sulfur. There are hundreds of different proteins, and they constitute at least half of the dry weight of living matter.

protist—Any of the unicellular organisms of the kingdom Protista, which includes protozoans, bacteria, some algae, and other forms not easily classified as plant or animal.

reactant—A starting material of a chemical reaction. The reactants are shown on the left side of the arrow in a chemical equation.

respiration—An oxygen-consuming metabolic process used by plants and animals to break down organic substances, yielding energy and releasing carbon dioxide.

species—A fundamental category in the classification of living organisms, ranking after a genus. Organisms of the same species share common characteristics and appearance, and they are capable of interbreeding.

squamous cell carcinoma—A relatively common skin cancer that can result from exposure to sunlight. It occurs in what is known as the Malpighian layer of the epidermis.

stomates—The tiny pores on the surfaces of plant leaves and stems, through which gases and water vapor are transferred between the atmosphere and the plant. The size of the opening varies depending on environmental conditions; it is crucial in controlling the amount of water lost from the interior of the plant.

transpiration—The evaporation of water from the interior of plants or animals. In plants, the size of the stomatal openings regulates this water loss. In the case of corn, about 100 pounds of water are transpired for every pound of dried plant material (leaves, stems, grain, roots, cobs).

ultraviolet radiation—Electromagnetic radiation of wavelengths between 100 and 400 nanometers. This range is somewhat arbitrarily divided into three segments, which increase in wavelength and decrease in energy: UV-C (100–280 nanometers), UV-B (280–320 nm), and UV-A (320–400 nm). Biological effects are greatest for UV-C and smallest for UV-A. UV radiation is higher in energy than visible light and is invisible to the human eye.

zooplankton—Animal life, often microscopic, that drifts in oceans or lakes.

RECOMMENDED READING

Beukema, J., W.J. Wolff, and J.J.W.M. Brouns. *Expected Effects of Climatic Change on Marine Coastal Ecosystems.* Dordrecht, The Netherlands: Kluwer Academic Publishers, 1990.

Bazzaz, F.A., and Fajer, E.D. Plant life in a CO_2-rich world. *Scientific American,* January 1992, 68–74.

Firor, John. *The Changing Atmosphere: A Global Challenge.* New Haven: Yale University Press, 1990.

Ford, M. *The Changing Climate: Responses of the Natural Fauna and Flora.* London: Allen and Unwin, 1982.

Gates, David M. *Climate Change and Its Biological Consequences.* Sunderland, Mass.: Sinauer, 1993.

Giese, A. *Living with Our Sun's Rays.* New York: Plenum Press, 1976.

Graham, R.L., M.G. Turner, and V.H. Dale. How increasing CO_2 and climate change affect forests. *Bioscience 40* (1990) No. 8, 575–87.

Jones, R.R., and T. Wigley. *Ozone Depletion: Health and Environmental Consequences.* Chichester, England: Wiley and Sons, 1989.

Kemp, D.D. *Global Environmental Issues: A Climatological Approach.* London: Routledge, 1990.

Krupa, S.V., and W.J. Manning. Atmospheric ozone: Formation and effects on vegetation. *Environmental Pollution 50* (1988) Nos. 1 & 2, 101–137.

MacKenzie, J.T., and M.T. El-Ashry. *Air Pollution's Toll on Forests and Crops.* New Haven: Yale University Press, 1989.

Peters, R., and J. Darling. The greenhouse effect and nature reserves. *Bioscience 35* (1985) No. 3, 707–17.

Roberts, L. How fast can trees migrate? *Science 243* (1989) No. 4892, 735–37.

United Nations Environment Programme. *Environmental Effects Panel Report.* Nairobi, Kenya: UNEP, 1989.

———. *Environmental Effects of Ozone Depletion: 1991 Update.* Nairobi, Kenya: UNEP, 1991.

Voytek, M. Addressing the biological effects of decreased ozone on the Antarctic environment. *Ambio 19* (1990) No. 2, 52–61.

World Meteorological Organization. *Scientific Assessment of Stratospheric Ozone: 1989.* Global Ozone Research and Monitoring Project Report No. 20. Washington, D.C.: National Aeronautics and Space Administration, 1990.

Wyman, R.L. *Global Climate Change and Life on Earth.* London: Chapman and Hall, 1991.

INDEX

air pollutants, 7
    carbon monoxide, 7
    lead, 7
    nitrogen oxides, 7
    ozone, 7
    sulfur dioxide. 7

biomes, 3, 4
    and climate, 4
    world distribution, 4
biosphere, 2, 3
    as a unit of organization, 2

carbon cycle, 13
carbon dioxide
    as a greenhouse gas, 9
    cycling, 9–14
carbon dioxide
    marine life, effects on, 19
    photosynthesis, effects on, 9–14, 31
    plant communities, effects on, 17–19, 31
    plant growth, effects on, 14–19, 31
    respiration, effects on, 9–14
carbon monoxide, 7
chemical environment,
    as a climatic factor, 6–7
chlorophyll, 11
climate, effect of
    greenhouse gases, 20
climatic factors, 5–8
    chemical environment, 6
    clouds, 8
    solar radiation, 8
    temperature, 5
    water, 6
clouds, 8
communities
    as a unit of organization, 2

    effects of carbon dioxide on, 17–19
    effects of temperature on, 21–25

decomposition
    and the carbon cycle, 13
DNA
    UV radiation, damage by, 26–28

ecosystem
    as a unit of organization, 2
    effects of climatic change, 43–46

global warming, effects on
    ecosystems, 21–25, 43–46
    human health, 25
    sea level, 25
greenhouse gases
    carbon dioxide, 9
    chlorofluorocarbons, 12
    direct effects of, 14–20
    indirect effects of, 20–25
    methane, 12
    nitrous oxide, 12
    ozone, 12

homeotherms, 5
human health, effects of
    global warming, 21
    ozone, 37
    UV radiation, 30
hydrogen peroxide
    as a tropospheric oxidant, 33–35
    in ice cores, 34

individual
    as a unit of organization, 2

lead, 7

INDEX

living systems
    biome, organization of, 2
    biosphere, organization of, 2
    and climate, distribution of, 3
    community, organization of, 2
    ecosystem, organization of, 2
    individual, organization of, 2
    population, organization of, 2
    solar radiation, effects of, 8
    and space, distribution of, 3
    species, organization of, 2
    and time, distribution of, 3
    wind, effects of, 8

marine life
    carbon dioxide, effects of, 19
    ocean currents, effect of, 8
    UV radiation, effects of, 32–33
methane, 12

nitrogen oxide
    as an air pollutant, 7
    as a greenhouse gas, 12

oceanic currents
    marine life, effects on, 8
ozone
    air pollutant, 7
    damage to plants, 36–37
    depletion, 26
    greenhouse gas, 12
    human health, effects on, 37
    tropospheric oxidant, 33

photosynthesis
    and carbon dioxide, 9–14
    and carbon cycling, 13
    and chlorophyll, 11
    temperature, effect of, 5
plant communities
    carbon dioxide, effects of, 17–19
plant growth
    carbon dioxide, effects of, 14–19
poikilotherms
    body temperature, 5

population
    as a unit of organization, 2

respiration
    and carbon cycling, 13
    and carbon dioxide, 9–14
    temperature, effect of, 5

sea level
    global warming, effect of, 25
solar radiation
    as a climate factor, 8
    living systems, effects on, 8
    photosynthetically active, 8
species
    as a unit of organization, 2
    temperature, effects of, 21–25

temperature
    enzyme function, effect on, 5
    global warming, 21–25
    as a climatic factor, 5
    of homeotherms, 5
    of poikilotherms, 5
    photosynthesis, effect on, 5
    respiration, effect on, 5
    species' distributions, effect on, 21–25
terrestrial animals
    UV radiation, effects of, 28–30
    skin cancers, 30
terrestrial plants
    UV radiation, effects on, 30–32
tropospheric oxidants, 33–37

ultraviolet radiation
    DNA, effects on, 26–28
    human health, effects on, 30
    marine life, effects on, 32–33
    terrestrial plants, effects on, 30–32
    terrestrial animals, effects on, 28–30

water, and living systems
    as a climatic factor, 6
    essential for life, 6
wind
    living systems, effects on, 8